Clinical Guide to Angio-OCT
Non Invasive, Dyeless OCT Angiography

Clinical Guide to Angio-OCT
Non Invasive, Dyeless OCT Angiography

Authors

Bruno Lumbroso MD
Director, Centro Oftalmologico
Mediterraneo for Retinal Diseases
Rome, Italy
Former Head of Department and
Director
Rome Eye Hospital, Rome, Italy

David Huang MD PhD
Professor
Department of Ophthalmology
Casey Eye Institute, Oregon
Health and Science University
Portland, Oregon, USA

Yali Jia PhD
Research Assistant Professor
Department of Ophthalmology
Casey Eye Institute, Oregon
Health and Science University
Portland, Oregon, USA

James G Fujimoto PhD
Professor of Electrical Engineering and
Computer Science
Department of Electrical Engineering and
Research Laboratory of Electronics
Massachusetts Institute of Technology
Cambridge, USA

Marco Rispoli MD
Staff Ophthalmologist
Department of Ophthalmology
Ospedale Nuovo Regina Margherita and
Centro Oftalmologico Mediterraneo for
Retinal Diseases
Rome, Italy

Contributors

Luca Di Antonio MD PhD, Jay S Duker MD PhD, Adil El Maftouhi OD, Woo Jhon Choi PhD, James G Fujimoto PhD,
David Huang MD PhD, Yali Jia PhD, Bruno Lumbroso MD, Leonardo Mastropasqua MD, Marco Rispoli MD,
Maria Cristina Savastano MD PhD, Giovanni Staurenghi MD, Nadia K Waheed MD MPH

JAYPEE *The Health Sciences Publishers*

New Delhi | London | Philadelphia | Panama

Jaypee Brothers Medical Publishers (P) Ltd.

Headquarters
Jaypee Brothers Medical Publishers (P) Ltd.
4838/24, Ansari Road, Daryaganj
New Delhi 110 002, India
Phone: +91-11-43574357
Fax: +91-11-43574314
E-mail: jaypee@jaypeebrothers.com

Overseas Offices

J.P. Medical Ltd.
83, Victoria Street, London
SW1H 0HW (UK)
Phone: +44-20 3170 8910
Fax: +44(0) 20 3008 6180
E-mail: info@jpmedpub.com

Jaypee-Highlights Medical Publishers Inc.
City of Knowledge, Bld. 237, Clayton
Panama City, Panama
Phone: +1 507-301-0496
Fax: +1 507-301-0499
E-mail: cservice@jphmedical.com

Jaypee Medical Inc.
The Bourse
111, South Independence Mall East
Suite 835, Philadelphia, PA 19106, USA
Phone: +1 267-519-9789
E-mail: jpmed.us@gmail.com

Jaypee Brothers Medical Publishers (P) Ltd.
17/1-B, Babar Road, Block-B, Shaymali
Mohammadpur, Dhaka-1207
Bangladesh
Mobile: +08801912003485
E-mail: jaypeedhaka@gmail.com

Jaypee Brothers Medical Publishers (P) Ltd.
Bhotahity, Kathmandu, Nepal
Phone: +977-9741283608
E-mail: Kathmandu@jaypeebrothers.com

Website: www.jaypeebrothers.com
Website: www.jaypeedigital.com

© 2015, Jaypee Brothers Medical Publishers

The views and opinions expressed in this book are solely those of the original contributor(s)/author(s) and do not necessarily represent those of editor(s) of the book.

All rights reserved. No part of this publication may be reproduced, stored or transmitted in any form or by any means, electronic, mechanical, photocopying, recording or otherwise, without the prior permission in writing of the publishers.

All brand names and product names used in this book are trade names, service marks, trademarks or registered trademarks of their respective owners. The publisher is not associated with any product or vendor mentioned in this book.

Medical knowledge and practice change constantly. This book is designed to provide accurate, authoritative information about the subject matter in question. However, readers are advised to check the most current information available on procedures included and check information from the manufacturer of each product to be administered, to verify the recommended dose, formula, method and duration of administration, adverse effects and contraindications. It is the responsibility of the practitioner to take all appropriate safety precautions. Neither the publisher nor the author(s)/editor(s) assume any liability for any injury and/or damage to persons or property arising from or related to use of material in this book.

This book is sold on the understanding that the publisher is not engaged in providing professional medical services. If such advice or services are required, the services of a competent medical professional should be sought.

Every effort has been made where necessary to contact holders of copyright to obtain permission to reproduce copyright material. If any have been inadvertently overlooked, the publisher will be pleased to make the necessary arrangements at the first opportunity.

Inquiries for bulk sales may be solicited at: jaypee@jaypeebrothers.com

Clinical Guide to Angio-OCT (Non Invasive, Dyeless OCT Angiography)

First Edition: **2015**

ISBN: 978-93-5152-399-4

Printed at: Ajanta Offset & Packagings Ltd., New Delhi

Contributors

Luca Di Antonio MD PhD
Retina Fellow
Ophthalmology Clinic
Department of Medicine and Aging Science
University G d'Annunzio of Chieti-Pescara
Italy

Yali Jia PhD
Research Assistant Professor
Department of Ophthalmology
Casey Eye Institute, Oregon
Health and Science University
Portland, Oregon, USA

Woo Jhon Choi PhD
Engineer
Department of Electrical Engineering and
Research Laboratory of Electronics
Massachusetts Institute of Technology
Cambridge, USA

Bruno Lumbroso MD
Director, Centro Oftalmologico Mediterraneo
for Retinal Diseases
Rome, Italy
Former Head of Department and
Director
Rome Eye Hospital, Rome, Italy

Jay S Duker MD PhD
Director
New England Eye Center
Professor and Chair of Ophthalmology
Tufts Medical Center
Tufts University School of Medicine
Boston, USA

Adil El Maftouhi OD
Orthoptist
Centre Explore Vision
Paris, France
Centre Hospitalier National Ophtalmologique
del XV XX
Service du Pr C Baudouin
Paris, France

James G Fujimoto PhD
Professor of Electrical Engineering and
Computer Science
Department of Electrical Engineering and
Research Laboratory of Electronics
Massachusetts Institute of Technology
Cambridge, USA

Leonardo Mastropasqua MD
Full Professor in Ophthalmology
Head, Department of Ophthalmology
University G d'Annunzio of
Chieti-Pescara, Center of Excellence
National High-Tech Center (CNAT)
Italian School of Robotic Surgery in
Ophthalmology, Italy

David Huang MD PhD
Professor
Department of Ophthalmology
Casey Eye Institute, Oregon
Health and Science University
Portland, Oregon, USA

Marco Rispoli MD
Staff Ophthalmologist
Department of Ophthalmology
Ospedale Nuovo Regina Margherita and
Centro Oftalmologico Mediterraneo for Retinal
Diseases
Rome, Italy

Maria Cristina Savastano MD PhD
Retina Fellow
School of Medicine
Catholic University
Rome, Italy

Nadia K Waheed MD MPH
New England Eye Center
Assistant Professor of Ophthalmology
Tufts Medical Center
Tufts University School of Medicine
Boston, USA

Giovanni Staurenghi MD
Professor of Ophthalmology
Chairman, Eye Clinic
Director, Residency Program
Department of Biomedical and Clinical
Sciences "Luigi Sacco"
Sacco Hospital
University of Milan, Italy

Preface

One of the most important present topics in medical retina is optical coherence tomography angiography (Angio-OCT). Certainly, the interest is growing almost exponentially, and it is discussed in all retina meetings. Several different Angio-OCT prototypes are developing, but only one device is clinically available such as XR-Avanti Optovue with split-spectrum amplitude-decorrelation angiography (SSADA) algorithm. SSADA technology was created and developed by David Huang and Yali Jia. We are seeing interesting results that can be applied immediately to everyday clinical ophthalmology, i.e. probably the latest and exciting development in the retina field. It is currently a very exciting field. This technology is evolving very fast and will continue to gain even more widespread applicability.

Dyeless Angio-OCT is fast, easy and non invasive. Next software will still improve retinal disorders and choroidal neovascularization (CNV) diagnostic, highlighting the membrane and easing clinical work.

The aim of the *Clinical Guide to Angio-OCT (Non Invasive, Dyeless OCT Angiography)* is to show everyday OCT user the clinical utility of Angio-OCT. The keyword is *Clinical*. The book presents the Angio-OCT principles. David Huang and Yali Jia present the SSADA algorithm and its applications, choroidal neovascularization (CNV), optic disk disorders and glaucoma. James G Fujimoto and Bruno Lumbroso propose the new terminology necessary to study this new technology. Clinical ophthalmologists, technicians and optometrists will then find a description of the vascular anatomy of retina and choroid, and an Angio-OCT Elementary Atlas of the general syndromes and clinical signs associations seen from vascular perspective. Then we present technological or scientific Angio-OCT progresses and try to understand their future clinical applications.

More work needs to be done to translate fully, the detail of Angio-OCT images into clinically useful information, and we still need to be cautious in interpreting the images and comparing them with histology data. We do not always have a significant idea of their true meaning. There is a debate from some specialists accustomed to fluorescein angiography, reluctant to adopt new technology.

The book discusses the clinical possibilities of Angio-OCT. It should guide the basic Ophthalmologist or Technician to understand the new imaging typical and atypical aspects. We hope to develop interest in using Angio-OCT in normal clinical work, show Angio-OCT users the practical clinical interest in day-to-day ophthalmology, and help them in understanding this new angiography.

Most imaging in the book (77 figures in Chapters 5, 7.1 and 10) has been performed with XR-Avanti Optovue SD-OCT device using SSADA algorithm. A few figures in chapters 3, 7.2 and 8 have been performed with a swept source OCT prototype using SSADA algorithm. Another technique is described in chapter 11, using ultrahigh speed, SS-OCT and its applications to optical coherence tomography-angiography (OCTA).

The book is fortunate to have outstanding contributions from scientific and clinical leaders in this field. We hope that it is a timely answer to a clinical need.

Bruno Lumbroso
Marco Rispoli

Acknowledgments

The authors gratefully acknowledge Donata Piccioli for the beautiful drawings. It is with immense gratitude that we acknowledge the support of many outstanding experts.

We express our gratitude to M/s Jaypee Brothers Medical Publishers (P) Ltd, New Delhi, India for their patience, encouragement and professionalism during the entire process. We are especially grateful to Shri Jitendar P Vij (Group Chairman), Mr Ankit Vij (Group President), Mr Tarun Duneja (Director-Publishing), Ms Samina Khan (PA to Director-Publishing), Mr Sunil Kumar Dogra (Production Executive), Mr Neelambar Pant (Production Coordinator), Mr Pawan Kumar (Graphic Designer), Mr Himanshu Sharma (Proofreader) and Mr Manver Singh (Typesetter) for shaping up of the book and making all the changes, without any complaints.

Contents

1. Principles of Optical Coherence Tomography-Angiography   1
Bruno Lumbroso, Marco Rispoli
Limitations of Optical Coherence Tomography-Angiography   *2*
Limitations of Optical Coherence Tomography   *2*
Retinal Fluorangiography and Optical Coherence Tomography-Angiography   *3*

2. Optical Coherence Tomography-Angiography: New Clinical Terminology   5
James G Fujimoto, Bruno Lumbroso, Marco Rispoli
Angio-OCT Clinical Terminology   *5*

3. Split-Spectrum Amplitude-Decorrelation Angiography   8
David Huang, Yali Jia

4. Solving the Practical Problems of Optical Coherence Tomography-Angiography   10
Marco Rispoli, Bruno Lumbroso
Mistakes to be Avoided   *10*

5. Optical Coherence Tomography-Angiography of a Normal Retina: The Anatomy of Blood Supply in the Retina   12
Maria Cristina Savastano, Bruno Lumbroso, Marco Rispoli
Posterior Pole   *12*
Macular Area   *12*
Arteries and Retinal Veins   *12*

6. Analysis and Synthesis: Analysis and Interpretation of a Pathological Optical Coherence Tomography-Angiography   16
Bruno Lumbroso, Marco Rispoli
Analytic Steps   *16*
Analysis of Neovascular Membranes   *21*
Synthesis   *22*

7.1 Clinical Applications: Aspects of OCT SSADA Angiography in Eye Disorders   23
Bruno Lumbroso, Marco Rispoli, Maria Cristina Savastano, Adil El Maftouhi, Leonardo Mastropasqua, Luca Di Antonio, Giovanni Staurenghi
Age-Related Retinal Anomalies   *23*
Retinal Anomalies and Coat's Disease   *23*
Retinal Anomalies in Angiomatosis   *24*
Retinal Superficial Anomalies in Macular Pucker   *25*
Macroaneurysm   *25*

Diabetic Patients without Retinopathy 30
Branch Retinal Vein Occlusions 30
Recent or Long-Lasting Retinal Ischemias 32
Diabetic Retinopathy 35
Proliferative Diabetic Retinopathy 45
CNV: Neovascular Membranes in ARMD 47
Neovascular Membranes in Myopic Eyes 48
Idiopathic Polypoidal Choroidal Vasculopathy 49
Geographic or Atrophic Macular Degeneration 59
Optic Disk Disorders 59

7.2 Clinical Applications: Optical Coherence Tomography-Angiography of Choroidal Neovascularization in Age-Related Macular Degeneration 60
Yali Jia, David Huang

8. Optical Coherence Tomography-Angiography of Optic Disk and Peripapillary Retinal Perfusion in Glaucoma 64
Yali Jia, David Huang

9. Fluorescein Angiography and Optical Coherence Tomography-Angiography: Advantages and Disadvantages 68
Bruno Lumbroso, Marco Rispoli
General Fluorangiography Advantages 68
General Fluorangiography Disadvantages 68
Fluorangiography Imaging Advantages 68
General Angio-OCT Disadvantages 69
General Advantages of Angio-OCT Imaging 69
Angio-OCT Imaging Advantages 69
Angio-OCT Disadvantages 69
Other Non Invasive Techniques 70

10. Reporting an Optical Coherence Tomography-Angiography 71
Marco Rispoli
Analytic Steps 71
Case Reports 71

11. Future Ultrahigh Speed Swept-Source OCT Technology and OCT-Angiography 75
Nadia K Waheed, Woo Jhon Choi, Jay S Duker, James G Fujimoto
Swept-Source Optical Coherence Tomography for Ultrahigh Speeds 75
Optical Coherence Tomography-Angiography 77
Imaging the Choriocapillaris 77
Optical Coherence Tomography-Angiography in Diabetes 79
Optical Coherence Tomography-Angiography in Dry Age-Related Macular Degeneration 81

Index 85

Introduction

In the last few years, structural and functional optical coherence tomography (OCT) technology has seen new and revolutionary developments. The most important of which is arguably an OCT angiography (Angio-OCT). Angio-OCT is already playing an important role in clinical ophthalmology as a new, non invasive and dyeless diagnostic tool, which serves as an adjunct to, or even a replacement for fluorescein and indocyanine green (ICG) angiographies. Angio-OCT is bringing multiple technical and clinical improvements in the study of retinal diseases, glaucoma and optic nerve disorders. It enables rapid, high-resolution visualization of vascular structure in three dimensions as well as ease of repeated imaging.

In the *Clinical Guide to Angio-OCT (Non Invasive, Dyeless OCT Angiography)*, we offer a step-by-step guide for interpreting clinical images and data acquired by Angio-OCT. In this book, we present a logical method for interpreting ophthalmic images. The first phase is analytic. The second phase combines elementary components to synthesize the data, allowing an accurate diagnosis and treatment decision. We also update OCT terminology in order to have a standardized approach for assessing Angio-OCT features. The book explains similarities and differences between this new imaging method, and the classical fluorescein and ICG angiographies. Very soon new advances in technology will further improve Angio-OCT imaging, making day-to-day clinical work easier. We trust the book will help ophthalmologists, residents, ophthalmic technicians and optometrists to understand and appreciate the new possibilities offered by the latest Angio-OCT imaging technologies.

Bruno Lumbroso
David Huang
Yali Jia
James G Fujimoto
Marco Rispoli

CHAPTER

1

Principles of Optical Coherence Tomography-Angiography

Bruno Lumbroso, Marco Rispoli

Angio-OCT or Optical coherence tomography (OCT)-angiography (Angio-OCT) is a new method of analysis based on high-resolution imaging techniques whereby the retinal and choroidal circulation can be visualized without the need to injection of any contrast agent. This new technology is therefore noninvasive, unlike fluorescein angiography that is currently still considered to be the gold standard of retinal vascular imaging. Angio-OCT is capable of detecting endoluminal flow at any time, is thus independent of time, and of administration of the contrast medium as instead occurs with fluorangiography.

Optical coherence tomography-angiography offers a precise visualization of intravasal flow without the dynamics produced by the dye. Even though this ensures an accurate visualization of the vessels, it does create the need to interpret the images identifying which ones now constitute the diagnostic parameters of vascular defects.

With the Angio-OCT vessels can be analyzed in daily clinical practice without injecting any dye. The future possibilities of the data gathered with this new technique are to be assessed carefully, studied and compared against fluorangiography, the method that has thus far provided important information.

While fluorangiography and indocyanine green (ICG) angiography make it possible to visualize the layers of the chorioretina, thanks to the injection of a dye, Angio-OCT visualizes the vascularization by using the blood flowing in the vessels as contrast medium. This new technique is used for diagnosis and follow-up of most retina disorders.

Angiography with fluorescein offers two-dimensional images on a single plane. *Angio-OCT instead is three-dimensional*; the scans are extracted from a cube and in general, they are parallel to Bruch's membrane or to the pigment epithelium. Angio-OCT is one of the most important developments of en-face OCT.

Fluorescein angiography is *dynamic*. Fluorangiography has a duration in time, with initial, intermediate and late stages. It therefore has a beginning and an end whereas *Angio-OCT is static*: there are no differences between the images at a given moment and the images taken an hour or a day later. Even if Angio-OCT is based on blood motion, the basic scenario, the retina, remains static.

Another important difference between fluorangiography and Angio-OCT is the *frequency* with which the tests may be carried out. Being an invasive procedure, fluorangiography cannot be performed too often as it may cause serious problems, even though this is a rather seldom occurrence. Angio-OCT instead can be performed every time an OCT examination is performed. All it needs is extend the examination by a few seconds, without this causing the patient any discomfort.

The mode of action of Angio-OCT is similar to an OCT examination, with the study of the retinal cube. The split-spectrum amplitude-decorrelation angiography (SSADA) angiographic examination currently lasts for 14 seconds, while the upcoming instruments will be providing responses in 4 seconds. The images of normal and pathologic retinal blood supply are highly reproducible. The instrument is easy to use and the OCT learning curve is very short.

Optical coherence tomography-angiography has proven useful in studying retinal structures affected by a variety of disorders:
- Diabetic retinopathy
- Acquired and congenital retinopathies
- Acute and chronic epithelial disorders
- Subretinal neovascularization

- Venous occlusions
- Intraretinal and preretinal neovascularization
- Macular edema of different origins
- Serous and hemorrhage detachments of the retina and pigmented epithelium
- Senile and juvenile macular degenerations
- Glaucoma and its evolution and treatment
- Optic disk disorders.

Optical coherence tomography-angiography allows us to highlight, measure and determine vascular lesions.

The biggest difference with respect to fluorescein angiography and ICG is that there is no leakage, staining or pooling. There are no differences between the early and later steps of the examination.

The images obtained can be:
- Analyzed
- Quantified
- Saved
- Compared with the findings of later examinations.

Optical coherence tomography-angiography is a reliable and sensitive examination. The images are reproducible by different operators; scans are performed rapidly, in a simple manner and above all they do not imply the administration of any contrast or dye. Since in a certain number of conditions Angio-OCT can replace fluorangiography, especially in the follow up of retinal edema of different origins, one must not underestimate the fact that it is a noninvasive examination to be preferred over invasive examinations that may lead in a small percentage of cases to side effects, complications and even to medicolegal problems.

Optical coherence tomography-angiography allows performing the examination in a matter of seconds, making a diagnosis immediately and deciding whether an intervention is required or the disorder should be kept under follow up; it allows us to quantify the lesion and carry out pharmacological studies.

LIMITATIONS OF OPTICAL COHERENCE TOMOGRAPHY-ANGIOGRAPHY

- Scans are currently performed only at the posterior pole.
- Sizes for the time being are 3 mm × 3 mm or 6 mm × 6 mm, but 12 mm × 12 mm will soon be possible.
- Good mydriasis is useful even though at times good images can be obtained with a reacting pupil. It is difficult to obtain images with a miotic pupil.

Table 1.1 Optical coherence tomography-angiography (Angio-OCT) is

- Simple (short learning curve)
- Rapid (5 seconds)
- Reliable
- Sensitive
- Reproducible
- Noninvasive-contactless: Harmless

- The dioptric media need to be transparent and the cornea needs to be well hydrated, with lacrimal film or artificial tears.
- Optical coherence tomography-angiography is performed with difficulty or may even be impossible in cases of:
  - Corneal opacities or corneal edema
  - Lens opacities
  - Opaque or hazy vitreous
  - Vitreal hemorrhage.

Images can be obtained even in the presence of opacity caused by cataracts if they are not in an advanced stage and even when intraocular buffer solutions are used, like for instance silicone oil that however do cause errors in the measurements taken (Table 1.1).

It is used to:
- Make diagnoses
- Assess the indications for an intervention
- Store images
- Quantify lesions, evaluate thicknesses, volume, extent of a lesion
- Assess the effects of treatment
- Track the evolution of postoperative course of an intravitreal injection or the postlaser course.

LIMITATIONS OF OPTICAL COHERENCE TOMOGRAPHY

- Mydriasis is required.
- Dioptric media need to be transparent.
- Exploration is restricted to the posterior pole.
- Abundant tear film is required.

Optical coherence tomography-angiography allows us to highlight, measure and determine vascular lesions.
- There is no leakage.
- There is no staining.
- There is no pooling
- There are no differences between the early or later steps of the procedure (Tables 1.2 to 1.7).

RETINAL FLUORANGIOGRAPHY AND OPTICAL COHERENCE TOMOGRAPHY-ANGIOGRAPHY

Ever since the early studies by Novotny and Alvis more than 50 years ago, retinal fluorangiography has been considered to be the best imaging method for assessing and studying the retinal-choroidal-vascular components of the eye.[1,2]

Even though fluorangiography makes it possible to assess important microvascular details, it requires an intravenous injection of a dye. Dyes at times may cause side effects, such as nausea, vomiting, allergic reactions, etc.[3] Moreover, fluorescein spreads through fenestrations of the choriocapillaris which makes it difficult to assess this vascular layer that is of fundamental importance for nutrition of outer retina. ICG angiography provides a good visualization of the choroidal anatomy in that it does enter into the extravascular spaces, and it visualizes the details of the choriocapillaris.[4-6]

With the introduction of the OCT in 1991, clinical practice has undergone considerable development. The use of the high-resolution spectral domain OCT (SD-OCT) has provided information that is comparable to a histological examination. In spite of the rapid evolution of imaging, even the most modern OCTs do not provide an adequate visualization of the choroid vessels. This limitation often imposes to ask the patient to undergo both an OCT examination and a fluorangiography in order to study such vascular disorders as diabetic retinopathy, age-related macular degeneration, retinal-vascular occlusion, etc.

The more recent and innovative OCTs make it possible to study the retina, the choriocapillaris and the choroid without dye injection. This method is made possible through a variety of accurate analysis systems.

We can breakdown these methods into two major groups: *phase-based* and *amplitude-based*. Some phase-based techniques include Angio-OCT, optical-microangiography, Doppler variance and phase variance. Among the amplitude-based there is scattering-OCT, speckle variance, correlation map and SAADA.

Fingler et al.[7] and Kim et al.[8] were the first to use the system called *phase-variance OCT* (*PV-OCT*) to show the retinal microvasculature. This method identifies the moving regions on the B-scan versus the neighboring static areas. Both in the retina and choroid the moving structures are blood vessels that differ from the other tissues that may be considered relatively static. An alternative method for capturing images of the retinal microvasculature is the *Doppler OCT* that measures the changes in position between successive scans.

Table 1.2 Difference between fluorangiography and optical coherence tomography-angiography (Angio-OCT)

- Fluorangiography is two-dimensional: single plane.
- Optical coherence tomography-angiography (Angio-OCT) is three-dimensional: several planes.

Table 1.3 Comparison between fluorangiography and optical coherence tomography-angiography (Angio-OCT)

- Dynamic fluorangiography
- Static Angio-OCT
- Angio-OCT is derived from en-face OCT

Abbreviations: Angio-OCT, optical coherence tomography-angiography; OCT, optical coherence tomography

Table 1.4 Advantages of dye leakage, staining and pooling

- They highlight vessel walls alterations, dye leakage, areas where the dye pools.
- Inconvenients of leakage, staining and pooling:
  - Early leakage prevents visualization of precise structure of the preretinal new vessels and CNV.
  - They conceal the true size of the blood column and vessel lumen.

Abbreviation: CNV, choroidal neovascularization

Table 1.5 Optical coherence tomography-angiography (Angio-OCT) used alone is useful in

- Vascular occlusions
- Diabetic retinopathy
- Preretinal neovascular proliferation
- Subretinal macular new vessels
- Vascular malformations
- Angiomatoses
- General, acquired and congenital retinopathies
- Glaucoma
- Optic disk disorders

Table 1.6 Optical coherence tomography-angiography (Angio-OCT) associated with fluorangiography is useful in

- Vascular occlusions
- Diabetic retinopathy
- Retinal and choroidal inflammations

Table 1.7 Currently, optical coherence tomography-angiography (Angio-OCT) does not provide useful information in

- Uveitis
- Choroiditis

This information is used to calculate the flow that is parallel to the imaging direction (called axial flow).[9]

All these methods have been implemented by using the OCT imaging system SD and swept-source versions. Many of these techniques have been analyzed both the transverse and axial images which provide a good visualization of the retinal and choroidal microvasculature. Even though all the methods provide qualitatively good images, the quantitative results capable of analyzing flow velocity are still somewhat limited.

The best method appears to be the SAADA for its potential use in measuring flow velocity in OCT.[10] Its potential advantage lies in the algorithm that creates a digital isotropic coherence volume before the correlation calculation. This procedure makes the algorithm equally sensitive both to movements along the longitudinal and transverse planes. SAADA may be used to quantify flows that are independent of the Doppler angle.

ACKNOWLEDGMENT

The authors gratefully acknowledge David Huang and Yali Jia, from the Casey Eye Institute, Portland, Oregon, who have developed SSADA algorithm. In this handbook, they have used SSADA algorithm Angio-OCT images.

REFERENCES

1. Alvis D. Happy 50th birthday. Ophthalmology. 2009;116(11): 2259.
2. Marmor MF, Ravin JG. Fluorescein angiography: insight and serendipity a half century ago. Arch Ophthalmol. 2011;129(7): 943-8.
3. Lipson BK, Yannuzzi LA. Complications of intravenous fluorescein injections. Int Ophthalmol Clin. 1989;29(3):200-5.
4. Owens SL. Indocyanine green angiography. Br J Ophthalmol. 1996;80(3):263-6.
5. Pauleikhoff D, Spital G, Radermacher M, Brumm GA, Lommatzsch A, Bird AC. A fluorescein and indocyanine green angiographic study of choriocapillaris in age-related macular disease. Arch Ophthalmol. 1999;117(10):1353-8.
6. Flower RW, Fryczkowski AW, McLeod DS. Variability in choriocapillaris blood flow distribution. Invest Ophthalmol Vis Sci. 1995;36(7):1247-58.
7. Fingler J, Schwartz D, Yang C, Fraser SE. Mobility and transverse flow visualization using phase variance contrast with spectral domain optical coherence tomography. Opt Express. 2007;15(20):12636-53.
8. Kim DY, Fingler J, Werner JS, Schwartz DM, Fraser SE, Zawadzki RJ. In vivo volumetric imaging of human retinal circulation with phase-variance optical coherence tomography. Biomed Opt Express. 2011;2(6):1504-13.
9. White B, Pierce M, Nassif N, Cense B, Park B, Tearney G, et al. In vivo dynamic human retinal blood flow imaging using ultra-high-speed spectral domain optical coherence tomography. Opt Express. 2003;11(25):3490-7.
10. Jia Y, Tan O, Tokayer J, Potsaid B, Wang Y, Liu JJ, et al. Split-spectrum amplitude-decorrelation angiography with optical coherence tomography. Opt Express. 2012;20(4):4710-25.

CHAPTER

2

Optical Coherence Tomography-Angiography: New Clinical Terminology

James G Fujimoto, Bruno Lumbroso, Marco Rispoli

INTRODUCTION

The first two generations optical coherence tomography (OCT), time-domain OCT and spectral domain OCT/Fourier-domain OCT, are structural OCT.

The third-generation OCT, OCT-angiography (Angio-OCT), is a functional OCT which will generate Angio-OCT. So the system we use in this handbook, XR-Avanti Optovue, is an Angio-OCT which uses the split-spectrum amplitude-decorrelation angiography (SSADA) algorithm to produce Angio-OCT. We have to use a new terminology to describe the techniques and the images that Angio-OCT offers. In the analytic description of an Angio-OCT report, both technical and clinical aspects must be taken into account.

ANGIO-OCT CLINICAL TERMINOLOGY

- Level (depth)
- Reflectance
- Flow, blood cell motion signal and decorrelation signal
- Morphology and architecture
- Texture

Level (Depth)

An essential element is the location of the Angio-OCT image and its relative depth in retina and choroid. The offset indicates level or depth. The segmentation levels are generally located at the inner limiting membrane (ILM), the inner plexiform layer (IPL) and the ideal pigmented epithelium [retinal pigment epithelium (RPE) ref section], or better still Bruch's membrane (BM).

While referencing to the exact level is easy for the healthy eye, in clinical practice the IPL is frequently deformed, thickened, or unrecognizable. The reference levels are usually the ILM, the ideal position of pigmented epithelium profile (RPEref), or BM.

Caution: Unambiguous interpretation of en-face OCT and Angio-OCT images at a given level may require a careful, comprehensive and global examination. Retinal pathology can cause displacement of retinal layers from their normal levels. In addition, when retinal pathology is present, image processing software errors, segmentation errors, can incorrectly identify reference retinal layers, so it may be necessary to manually correct the segmentation in order to use correct reference boundaries.

Correct interpretation may require examination of en-face images at multiple depth levels and cross-sectional images through the region of pathology in order to confirm clinical interpretation of an en-face image at a specific level.

Reflectance

Reflectance is the contrast for structural OCT. In structural OCT, tissue density is shown by variation in reflectance.

Flow, Blood Cell Motion Signal, and Decorrelation Signal

Flow or blood cell motion creates contrast in Angio-OCT, and this motion contrast is different from reflectance as seen in structural OCT. Motion contrast is measured by the decorrelation signal.

This is another new and fundamental concept in the description of Angio-OCT. Angio-OCT generates motion contrast by performing repeated B-scans at the same retinal position. Volumetric three-dimensional Angio-OCT data is then generated by performing many such repeated B-scans covering an area of the fundus. Angio-OCT requires larger numbers of B-scans than structural OCT, therefore the scan time is longer or retinal area imaged is smaller. The repeated B-scans are compared numerically and a decorrelation signal is calculated which shows how much structures change between the repeated B-scans. Flow or blood cell motion causes changes between B-scans, and the decorrelation signal measures this motion contrast.

Optical coherence tomography-angiography has the advantage that it can visualize blood vessels and blood flow which may be transverse or perpendicular to the OCT beam. Doppler OCT cannot visualize perpendicular blood flow.

Vascular decorrelation signal: Flow or blood cell motion generates a decorrelation signal which produces contrast and enables visualization of three-dimensional retinal and choroidal vasculature.

Nonvascular decorrelation signal: Decorrelation signals can be generated by structures which are not moving blood. There are conditions where the decorrelation signal shows nonvascular reflecting structures that are present in the corresponding en-face structural imaging, for instance, hard exudates, pigment accumulation, thrombosed aneurysms and retinal hemorrhages. The decorrelation signals are associated with pathology that has very fine or inhomogeneous internal structure, as in the examples described.

This may be the result of very small eye motions or OCT scanning changes which cause these small structures to change in repeated B-scan images and generate a decorrelation signal. Larger or more homogenous structures must move much more in order to generate a decorrelation signal.

Flow velocity and decorrelation signal: The decorrelation signal has a *limited dynamic range* response to flow. Therefore, one must be careful interpreting the decorrelation signal directly as a flow velocity.

There is a *sensitivity limit* for the slowest flow that can be detected which is determined by the time between repeated B-scans. If the time between B-scans is increased, then the decorrelation signal becomes more sensitive to slower flows. However, increasing the time between B-scans also means the decorrelation signal and OCT-angiogram becomes more sensitive to eye motion and has more noise.

There is also a *saturation limit* where the decorrelation signal cannot differentiate faster flows, because any flow which is faster will produce the same decorrelation signal. This means that the decorrelation signal does not respond to changing flow velocity above this *saturation limit*.

For larger vessels, the flows are always fast and generate the same decorrelation signal or OCT angiogram. In other words, the flow is faster than the *saturation limit* and decorrelation signal is saturated.

However, for small vessels such as capillaries or the choriocapillaris, the flow may be too slow and below the *sensitivity limit*. The small vessel with slow flow may not generate a decorrelation signal and may be invisible on the OCT-angiogram, even though the vessel is present.

Therefore, if vascular features are not visible on Angio-OCT, it may mean that there is dropout or atrophy of the vasculature or reduced/impaired flow.

Morphology and Network Architecture

Shape and course of the vessels of the vascular networks: Vessel shape and vessel course are two other parameters to be assessed in Angio-OCT. Shape may be regular or irregular, while course may be regularly winding, stretched out, distorted and contorted. These features may coexist in various combinations in the various pathological presentations.

We should study vascular networks in Angio-OCT separately according to retinal layers.

A very important feature not detectable in traditional angiography is the presence of normal anastomoses between superficial vascular plexus and deep vascular plexus, defined as connections. There may also be anomalous pathological connections between superficial plexus and deep vascular plexus: we define them as shunts.

When studying the architecture, we evaluate the shape of the vessels that may be regular or irregular, and the density of the capillaries that may be sparse or dense. The cross-section of the capillaries shows their diameter that may be thin, large, regular or irregular. The architecture of the vascular network may be regular or irregular.

Architecture and Density of the Network

The *architecture* of the Angio-OCT in each retinal layer is assessed separately.

The caliber of the vessels is evaluated. It may be large or thin, irregular, and with or without dilatations.

The vascular network may be dense, sparse, widened, dilated, rarefied, weak, dense and tangled. The mesh may be sharp, distinct or indistinct.

Texture

Texture is a new concept in the description of Angio-OCT. Texture may be coarse, granulated, fine, faint, speckled and grayish. Texture is a very interesting concept. However, we should be careful assessing, because texture can change with image averaging. For example, fine texture in retinal layers changes depending on the amount of image averaging. Coarse texture is less sensitive to averaging effects. Since different softwares or instruments in the future may use different amounts of averaging, texture should be defined so that it works with future instruments (i.e. use features that are not so sensitive to image averaging).

- *Ischemia* shows a total disappearance of the capillaries, the area has a grayish aspect and the texture is more or less fine or finely granulated. The density of the collateral branches depends on the extent of the ischemia. Connections (shunts) between the retinal vascular superficial and deep plexuses are almost invisible in normal conditions. They are instead, extremely frequent and evident in the vascular pathology of the retina.
- *Edema* texture is a grayish area related to the widening of the vascular network meshes and rarefaction of the finer capillaries.

The texture should be assessed layer-by-layer and is related to the presence or absence of the vascular plexus that may be altered or absent.

Optical Coherence Tomography-angiography Terminology for Neovascular Membranes

Location (depth, offset): Location is the first point that needs to process because it is necessary to pinpoint the position of the neovascular membrane in the preretinal, intraretinal or subretinal, macular, or parapapillary context. It needs to be located with respect to BM. Age-related macular degeneration: in this disorder new vessels are detected mainly just below the pigmented epithelium, above BM (type I) and in fewer cases above RPE (type II).

Profile: The next step is assessing the profile of the membrane in order to decide on the best segmentation. The correct segmentation of a neovascular membrane does not always coincide with a correct segmentation of the surrounding retina.

Segmentation thickness: It is a necessary feature for assessing the planes of the neovascular membrane. The new vessels may vary considerably in thickness because they develop in a random manner and not within the normal retinal layers.

The *thickness of the segmentation* is of great importance in relation to the average size of the new vessels. If the new vessels are rather thin, a thick segmentation could produce a confused image of the flows on several planes. If the new vessels are rather thick, a thin segmentation will produce a wrong image of the choroidal neovascularization (CNV) flows.

Choroidal neovascularization morphology: In naïve patients, new vessels appear in the form of branched structures with evidence of a main afferent trunk, or they may have a roundish aspect, like a bicycle wheel. Peripheral anastomoses are evident, and they constitute an important parameter to check in the follow up.

These flows have a clear and evident aspect, and it is possible to follow the course of the main and secondary branches. The neovascular membranes flow appears to be more evident than normal adjacent flows.

In case of neovascular membranes treated with anti-vascular endothelial growth factor (VEGF), CNV network appears smaller, with less anastomoses, but above all, after treatment the changes in the flow produce a *fragmented* aspect of the network. It is disaggregated, making it difficult to follow the smaller branches course. The secondary branches decrease or disappear immediately after injection and reappear after 2 or 3 weeks.

CHAPTER

3

Split-Spectrum Amplitude-Decorrelation Angiography

David Huang, Yali Jia

INTRODUCTION

Optical coherence tomography (OCT) is part of the standard of care for retinal diseases, and increasingly used for the management of glaucoma and anterior segment surgeries as well. However, conventional structural OCT cannot provide direct blood flow information. Since 2007, several OCT angiography methods have been described for three-dimensional (3D) noninvasive vasculature mapping at the microcirculation level. Our group recently developed the split-spectrum amplitude-decorrelation angiography (SSADA) algorithm to improve the signal-to-noise ratio (SNR) of flow detection.[1] The theory of SSADA is described below.

Optical coherence tomography angiography uses flow as the intrinsic contrast; therefore no injection of an extrinsic contrast agent is needed. Flow is detected as a variation over time in the speckle pattern formed by interference of light scattered from blood cells and adjacent tissue structure. Several techniques have been developed to detect flow based on variations in the magnitude or phase of the OCT signal in the same location over consecutive image frames. These techniques are sensitive to flow in both transverse and axial directions. This differs from Doppler OCT, which only measures axial velocity. Our SSADA algorithm is based on amplitude decorrelation and does not use phase information. Amplitude (magnitude) information is more reliable than phase information, which can be degraded by noise from OCT system phase noise and background tissue motion. Both decorrelation and variance functions provide information on variation. We chose to use decorrelation because it is not affected by the average signal strength.

In the SSADA technique, OCT signal is first split into several (four or more) spectral bands. So instead of a single image frame with high axial resolution, several low-resolution images are obtained. Lower axial resolution means a wider coherence gate over which reflected signal from a moving blood cell can interfere with adjacent structures, thereby increasing speckle contrast. Furthermore, each spectral band forms a different speckle pattern, providing independent information on flow. When amplitude decorrelation images from the multiple spectral bands are summed together, the flow signal is increased. The lower axial resolution of split-spectrum OCT also decreases noise from axial eye motion driven by the retrobulbar pulsation. By enhancing flow signal and suppressing bulk motion noise, SSADA improves the SNR of flow detection by at least a factor of two,[1] depending on the number of spectral bands employed.

Split-spectrum amplitude-decorrelation angiography can work with any high-speed OCT and that all of the Fourier-domain OCT systems, including spectral domain OCT and swept-source OCT (SS-OCT) systems. Higher speed and lower amplitude noise allows for larger angiograms. SS-OCT at 1,050 nm provides better detail of outer choroid due to better penetration and less washout of interference fringes from high flow velocity. Spectral OCT at 840 nm appears to have lower amplitude noise and can provide up to 6 mm angiograms at just 70 kHz axial repetition rate when using the highly efficient SSADA algorithm. As shown by an example from en face angiograms of the macular retinal circulation with 70 kHz 840 nm spectral OCT (Figs 3.1A to H), SSADA provides a clean and continuous microvascular network and less noise just inside the foveal avascular zone (FAZ). We found that the novel SSADA algorithm improved both the SNR of flow detection and the connectivity of the imaged vascular network by more than a factor of two when compared to the full-spectrum algorithm (Fig. 3.1B). This is equivalent

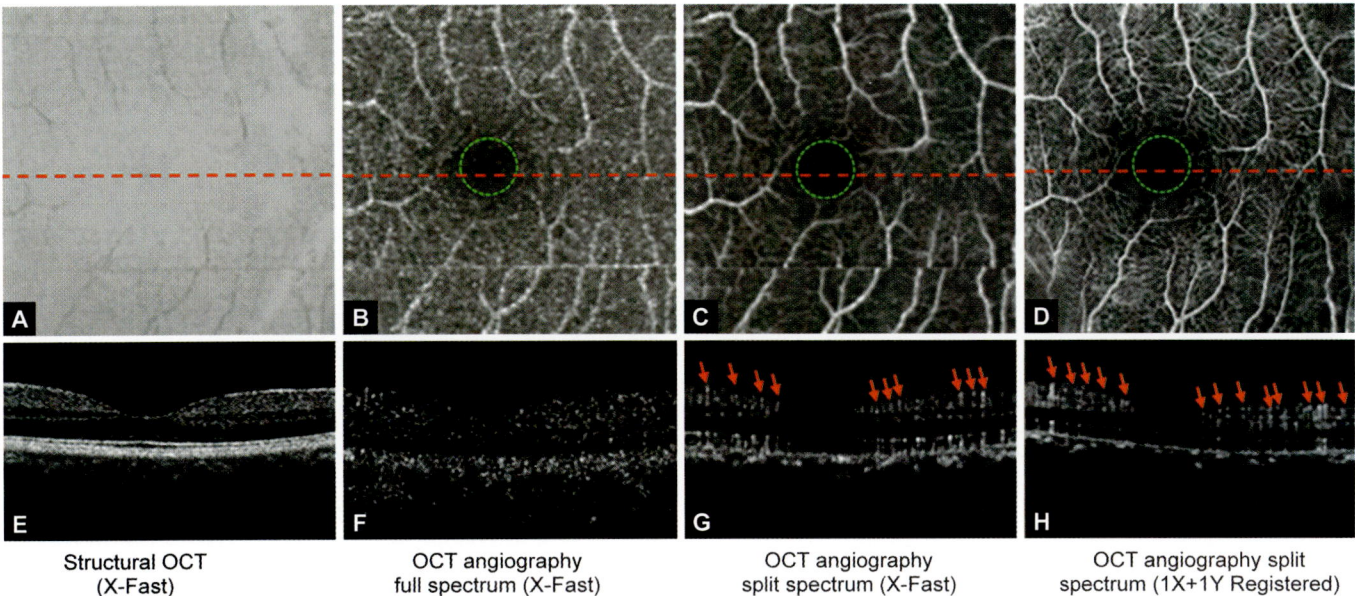

| Structural OCT (X-Fast) | OCT angiography full spectrum (X-Fast) | OCT angiography split spectrum (X-Fast) | OCT angiography split spectrum (1X+1Y Registered) |

Figs 3.1A to H Comparison of structural optical coherence tomography (OCT) (A, E) and amplitude-decorrelation angiograms of the macula (3 mm × 3 mm area) using full-spectrum (B, F), split-spectrum (C, G) and split-spectrum averaged angiograms from 1X-fast and 1Y-fast scans after three-dimensional (3D) registration (D, H). En face maximum decorrelation projections of retinal circulation showed less noise inside the foveal avascular zone (FAZ, inside green-dotted circles) and more continuous perifoveal vascular networks using the novel split-spectrum algorithm (C) compared to standard full-spectrum algorithm (B). The cross-sectional angiograms (scanned across dashed line in B and C) showed more clearly delineated retinal vessels (red arrows in G) and less noise using the split-spectrum algorithm (G) compared to the standard (F). There are saccadic motion artifacts that appear as artifactual horizontal lines in B and C. This and other motion artifacts are removed using the 3D registration algorithm, resulting in a continuous artifact-free microvascular network in D. The registration and averaging of two orthogonal scans also removed motion blur and further improved signal-to-noise ratio (SNR), allowing the visualization of a greater number of distinct small retinal vessels (microvascular network in D, red arrows in H).

to a speed gain of four times—meaning that SSADA is able to produce high-quality angiograms of useful dimensions (3 mm × 3 mm to 6 mm × 6 mm) using a commercially available 70 kHz OCT, while a simpler full-spectrum algorithm would require a 280 kHz OCT, which is not commercially available.

Besides its noninvasive nature, SSADA has several compelling characteristics that make it a promising modality for clinical use. SSADA can be acquired in a few seconds, compared to several minutes for fluorescein angiography (FA). Its 3D imaging allows for depth resolution of pathology and separation of individual vascular layers for evaluation.

Quantitative information, such as vessel density, vessel area and flow index, can now be obtained. The scan pattern and SSADA processing can be implemented on spectral-domain or SS-OCT systems without any special hardware modification.

REFERENCE

1. Jia Y, Tan O, Tokayer J, Potsaid B, Wang Y, Liu JJ, et al. Split-spectrum amplitude-decorrelation angiography with optical coherence tomography. Opt Express. 2012;20(4):4710-25.

CHAPTER

4

Solving the Practical Problems of Optical Coherence Tomography-Angiography

Marco Rispoli, Bruno Lumbroso

INTRODUCTION

Angio-OCT offers satisfactory results and, at times enthusiastic results, but there are, initially, difficulties in the set-up that need to be understood in order to avoid evitable disappointments.

MISTAKES TO BE AVOIDED

Initially some operators use techniques that are not appropriate and that do not give good results.
Common technical criticisms:
- Using a 6 × 6 field gives poor resolution
- Full thickness modality, provides frequently confusing data, given the overlap of the layers that instead could be separated for easier readability. The full thickness modality does not provide data of the layers beneath the retinal pigment epithelium (RPE).

Optical coherence tomography-angiography without contrast agent presents essentially two limits:
1. Surface artefacts
2. Mirror effect of the RPE.

The first is currently being solved, while the second is linked to the algorithm; hence the solution is neither easy nor is it quick.

The difficulties about image capture time may be solved through the following actions:
- Decrease passages from 5 to 3 for each B-scan
- Carry out a single horizontal stack and a single vertical stack instead of two.
This should reduce the capture time by over one-third.

Some operators require the fixation dot to be mobile, even though probably nobody has noticed that the fixation is immobile, but the capture area can be shifted.

Often there is also an error in the segmentation of the RPE fit that does not follow the ideal profile but the real profile.

Changing the aiming dot shape seems to be an important thing. The request to change it has been made but the problem is not easy to solve.
- *Errors to be avoided when comparing the fluorescein angiography (FAG) and indocyanine green (ICG) angiography images with the Angio-OCT images:* One must not directly compare the Angio-OCT images with the FAG and ICG angiography images. Fluorangiography is equivalent to a full thickness Angio-OCT where all the layers are represented in a confused manner. The Angio-OCT gives good results if the layers are studied separately. There is poor concordance or inverse concordance as regards to hyperreflectivity/hyperfluorescence.

Some operators probably do not take into account the fact that the Angio-OCT is not absolutely a fluorangiography and hence one must not make a direct comparison between the images of the two techniques, but rather make a comparison between the representation of known elementary lesions (known through a traditional or en-face B-scan analyses).

For instance, microaneurysm:
- *Fluorescein angiography:* Hyperfluorescence due to accumulation, staining and leakage of the dye. Exclusive data of the technique: Location (XY, hence two-dimensional) and functional continence of the lesion (the greater the diffusion, the greater the imperfection of the wall).
- *Limits:* It does not provide figures on the real size that appears to be altered by the take-up and diffusion phenomena. No data on its location in the context of the retinal thickness (XY + Z), two-dimensional examination.

- *Optical coherence tomography-angiography:* The microaneurysm is not evident in 100% of cases. In order to be seen and studied, the lesion must be perfused and detectable with the technique. Initially located in the B-scan, thereafter, in order to get a good Angio-OCT image, we will have to find the best segmentation of the layer of interest and finally we will have to decide the thickness of the scan that will have to be equal to or greater than the size of the aneurysm. If these requirements are not complied with the representation of the lesion is altered or even absent.

From the above remarks, it is evident that a direct comparison between the two techniques is absolutely not feasible. There are some aspects that need to be dealt with carefully before matching the data.

- *Optical coherence tomography-angiography is a three-dimensional examination. Fluorangiography is two-dimensional:* The term "angiostratigraphy" means that we study two-dimensional sections taken out of a three-dimensional structure. Hence knowledge of the third dimension is required in order to be able to extract the Angio-OCT image that is to be juxtaposed to the FAG.

Using the split-spectrum amplitude-decorrelation angiography (SSADA) with this device has enabled us not to notice the negative signal-to-noise ratio, in that the images have been "cleared" of useless data (Table 4.1).

Table 4.1 Suggested settings for routine examinations

- *Global examination:*
  - Full thickness mode
  - From the ILM to the RPE
- *Superficial vascular network:*
  - *Segmentation:* ILM
  - *Thickness:* 60 μm
  - *Offset:* 6 μm
- *Deep vascular network:*
  - *Segmentation:* IPL
  - *Thickness:* 30 μm
  - *Offset:* OPL level
- *CNV:*
  - *Segmentation:* RPE
  - *Thickness:* 60–200 micron
  - *Offset:* RPE level

Abbreviations: ILM, internal limiting membrane; RPE, retinal pigment epithelium; IPL, inner plexiform layer; OPL, outer plexiform layer

A direct comparison between Angio-OCT and FA is not possible. The two imaging methods are complementary.

SUMMARY

In summary:
- Do not use 6 × 6 images to highlight lesions that require a better resolution (3 × 3 is more likely)
- Do not use a huge amount of retino- and angiographic data in an attempt to find a perfect match.

CHAPTER

5

Optical Coherence Tomography-Angiography of a Normal Retina: The Anatomy of Blood Supply in the Retina

Maria Cristina Savastano, Bruno Lumbroso, Marco Rispoli

INTRODUCTION

Angio-OCT gives the opportunity to see the histological vascular structure of the retina in vivo without contrast agent. In the past, this could be seen only in histological sections.

The periphery beyond the arcades cannot yet be examined with Angio-OCT. Fluorangiography requires the operator to have excellent knowledge of the anatomy of the retina. The study of Angio-OCT demands indepth knowledge of the histology. The transition from fluorangiography to Angio-OCT therefore implies a qualitative change in the way images are looked at.

The study of anatomy of the retina using OCT and Angio-OCT implies that the posterior pole can be broken down into the area lying inside the vascular arches and the retinal periphery that lies beyond these structures and the optic disk.

POSTERIOR POLE

The posterior pole, which is oval in shape with a long horizontal axis, measures around 8–10 mm (30–35).

The field covered by Angio-OCT, which is currently an area measuring 3 mm × 3 mm or 6 mm × 6 mm, does not allow us at the present time to study the whole of the posterior pole with a single cube up to the optic disk and to map the area. Very soon the field will be enlarged to 12 mm × 12 mm.

MACULAR AREA

The macula has a thickness of 160–190 μm and is located at the center of the posterior pole where it forms a slight depression containing the fovea and foveola; it has a diameter of 1,200 μm. Inside the macular area, angiography shows that there is avascular area having a normal diameter of 450–500 μm. This area is wider in case of diabetic retinopathy or of other ischemic retinopathies.

The fovea, where the concentration of cones is highest, is contained within a concentric circle, 350 μm in diameter inside the avascular angiographic area, while the foveola that corresponds to the macular center measures around 120–150 μm. In the macula, we find a high density of xanthophyllic pigment and the pigment epithelial cells here are the most densely packed and contain a high density of pigment granules. The interpapillomacular zone is a little thicker than other macular retina.

ARTERIES AND RETINAL VEINS

The optic papilla has a diameter of 1,500 μm while the retinal veins at the edge of the disk have a maximum diameter of around 120 μm. In the midperiphery, the veins have an average diameter of 60 μm. The retinal arteries have a smaller diameter: 80 μm at the edge of the disk, and 50 μm in midperiphery.

In contact with the retinal vessels, in the periarterial avascular area, the capillaries are very rare, virtually absent. Sizes of the arterial and venous capillaries of the retina range between 5 μm and 10 μm.

The sensory retina is supplied with two clearly distinct systems.

Retinal Vascular Networks

Analysis of Retinal Vascular Plexuses with Angio-OCT

The resolution of Angio-OCT split-spectrum amplitude-decorrelation angiography (SSADA) makes it possible to clearly visualize the superficial vascular plexus with a 60 μm section at the internal limiting membrane (ILM). The two deep plexuses (DPs) cannot be clearly differentiated, since the smallest of 30 μm structures do not have sufficient resolution for them to be clinically useful. These two plexuses will therefore be treated as a single vascular entity, included in a segmentation at the inner plexiform layer (IPL) of at least 30 μm.

- *Superficial vascular plexus:* Located in the ganglion cell layer and in the nerve fiber layer.
- *Deep vascular plexus:* Located in the inner nuclear layers and external plexiform. From the anatomical standpoint, this plexus consists of two additional plexuses located, respectively, on the inside of the inner nuclear and on the outside of the outer plexiform layer (OPL). They cannot be individually seen by the Angio-OCT, and therefore, they are considered to be a single plexus.

In order to study the two vascular plexuses we have used specific parameters that concern the intraretinal level (ILM, IPL, RPE, RPEref), the thickness of the scan being examined and the offset. The need to establish very precise points of reference to analyze the vascular plexuses is determined in order to compare the analyses made by different operators and hence make the images as objective as possible.

Figure 5.1 presents an OCT B-scan with the precise location of the superficial and deep vascular plexus. The superficial plexus (SP) is represented by the large retinal vessels located in the innermost layers whose measurement corresponds on an average to 120 μm. The DP extends between the innermost portion and outer portion of the OPL that measures on an average around 60 μm. The assessment of the SP uses a thickness of 60 μm from the ILM so as to include all the vessels of this plexus. The parameters for the DP are defined with reference to the IPL in a 30 μm thick scan to visualize the DP in its entirety (Figs 5.2A and B).

Optical coherence tomography-angiography shows different morphological features of the retinal blood supply for the two plexuses considered.

1. *The superficial plexus:* The vascular distribution is represented by multiple white linear structures (flow) against a black background that, with a centripetal pattern, converge towards the fovea and originate from the large upper and lower vascular arcades. Secondary vessels leave the main vessels, forming a web. The thickness of the vessels is homogeneous throughout the length of the scan. The course of the vessels is always rather linear; the web is regular without sudden changes in direction or without vascular meanders or loops. The vessels show a vascular

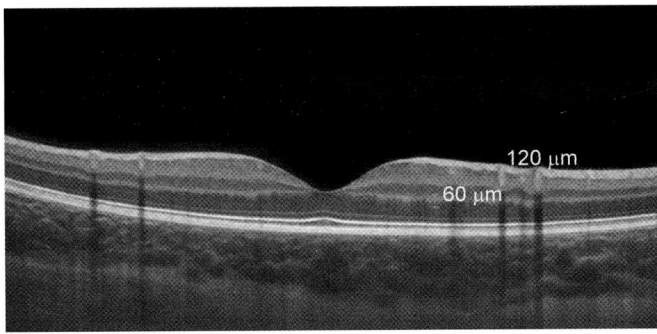

Fig. 5.1 Optical coherence tomography (OCT) B-scan with location of the superficial and deep vascular plexus. The superficial plexus (SP) is represented by large retinal vessels located in the innermost layers whose measurement corresponds on an average to 120 μm. The deep plexus (DP) extends between the innermost and outer portion of the outer plexiform layer (OPL) that measures on an average about 60 μm

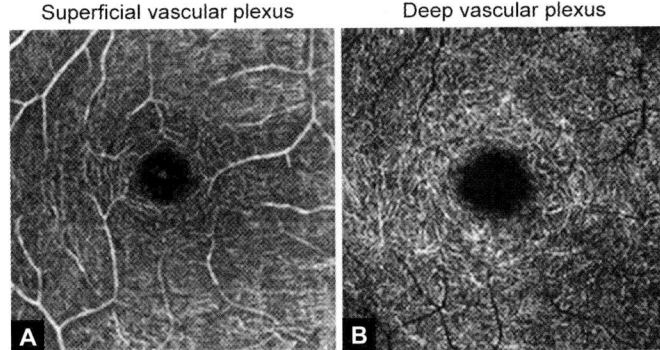

Figs 5.2A and B Optical coherence tomography (OCT) B-scan: (A) The evaluation of the superficial plexus (SP) is selected at a thickness of 60 μm from the internal limiting membrane (ILM) so as to include all the vessels of this plexus; (B) The parameters for the deep plexus (DP) are defined with reference to the inner plexiform layer (IPL) in a 30 μm thick scan in order to visualize the DP in its entirety

signal (decorrelation) throughout the scan. Around the avascular area, the capillaries form continuous perifoveal arcades with regular meshes (Fig. 5.3).

2. *The deep plexus:* It consists of a close-knit pattern of vessels whose orderly distribution around the avascular foveal zone presents numerous thin horizontal and radial interconnections. The pattern is concentric around the avascular foveal zone. Thickness of the vessels is constant throughout the scan as is their vascular signal. The vascular network consists of small vertical interconnection anastomoses between the superficial and deeper vessels of the same DP. From each lower extremity of the vertical anastomoses horizontal vessels fan out that interconnect to form a complex pattern (Figs 5.4 and 5.5).

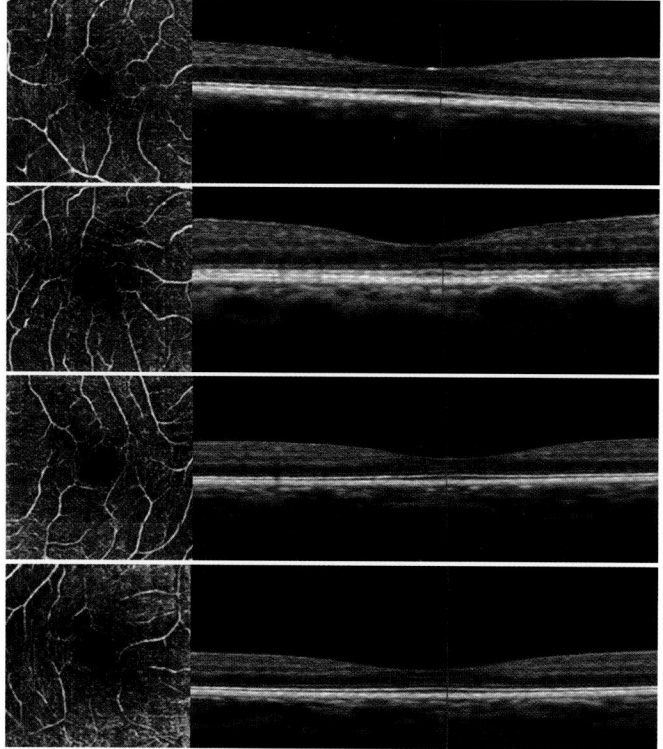

Fig. 5.3 Segmentation: internal limiting membrane (ILM); thickness: 60 μm; offset: 6 μm

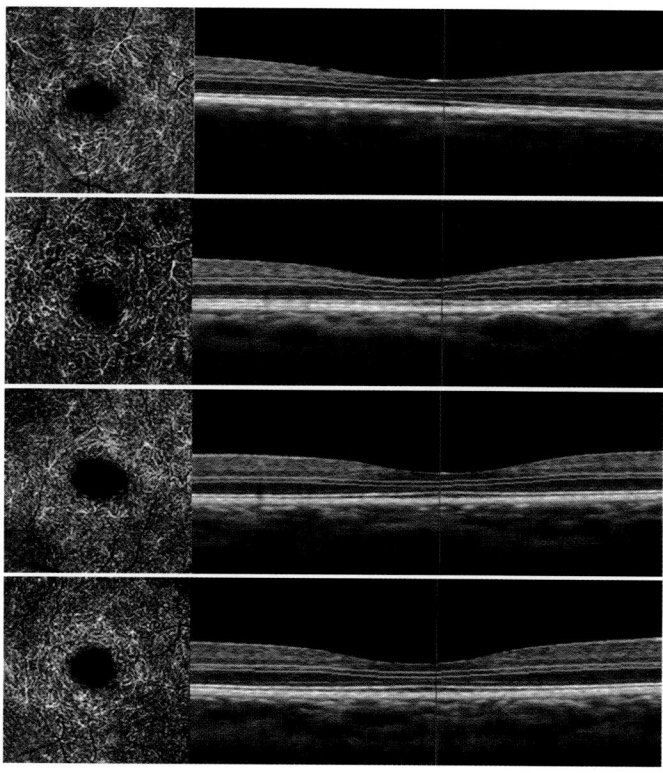

Fig. 5.4 Segmentation: inner plexiform layer (IPL); thickness: 30 μm; offset: outer plexiform layer (OPL) level

Optical coherence tomography-angiography can detect endoluminal flow at any time and is thus independent of the time of administration of the contrast medium as instead occurs with fluorescein angiography.

The classical anatomic studies carried out in the first half of the 20th century show that the distribution of retinal vessels is organized into three distinct layers: (1) SP, observable with the ophthalmoscope with the large- and average-sized vessels distributed in the retinal nerve fiber layer; (2) inner plexus, a body of small-sized capillaries located close to the inner surface of the internal nuclear layer; (3) outer plexus: morphologically similar to the internal plexus but located on the outer surface of the external plexiform layer.

Optical coherence tomography-angiography has confirmed these studies in vivo and allows us to study separately the two vascular plexuses, the superficial vascular plexus and the complex internal/external complex that we have considered as a single DP (Fig. 5.6). The two plexuses clearly have different features that cannot be distinguished by classical fluorescein angiography.

In fluorescein angiography, both plexuses overlap and therefore they cannot be distinguished nor assessed separately. The contemporary visualization of both plexuses

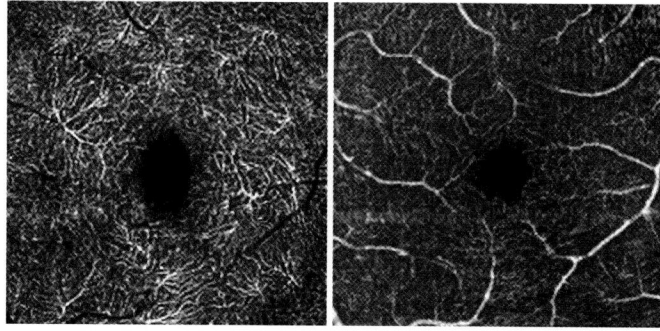

Fig. 5.5 Normal retina: Superficial vascular plexus; superficial retinal vessels show a centripetal distribution, originating from the vascular arches and converging towards the fovea. Some secondary vessels arise laterally from the main vessels forming a regular spider-web. The avascular area in the center can be noticed. The deep vascular plexus is concentric around the avascular fovea. The deep vessels arise from the outer extremity of the vertical anastomoses between superficial vessels and deep vessels. From the lower end sprout horizontal vascular fans that are interconnected and form a complex pattern

does not make it possible to analyze the superficial and deep vascular features that may be involved individually or separately in the various pathological processes. In healthy eyes, the SP consists of larger vessels with respect to the

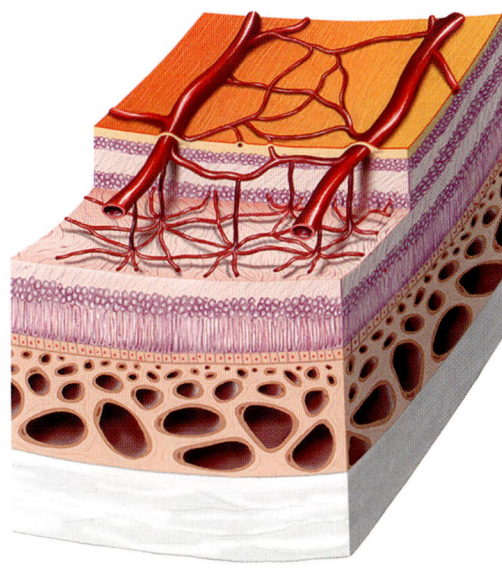

Fig. 5.6 Anatomy of the normal retina; optical coherence tomography-angiography (Angio-OCT) allows to breakdown blood supply to the normal retina into two networks: the superficial plexus (SP) and the deep plexus (DP)

deep complex: both plexuses are distributed according to a centripetal pattern around the avascular foveolar zone. The DP consists of small fan-shaped vessels that interconnect to form a complex pattern.

With fluorescein angiography the vascular tree and therefore the vessel morphology cannot be directly visualized; what is seen is the result of intravascular fluorescein. This effect is particularly evident in the case of retinal wall lesions and hence the diffusion (leakage) or accumulation masks the real aspect of the vessels.

Optical coherence tomography-angiography shows precisely the intravascular flow without dye. This warrants a good visualization of the vessels; however, it does entail new way to imaging interpretation. We need new diagnostic parameters for vascular disorders.

One of the important Angio-OCT limitations is the size of the scan area that involves only the macular region (3 mm × 3 mm to 6 mm × 6 mm). In the near future, a full field will be obtained that will provide further information and offer greater details. Another problem is the limitation in visualizing the structures under the RPE using the present softwares.

1. *Superficial plexus:* The vascular distribution is represented by many linear structures that converge towards the fovea and originate from the great upper and lower vascular arches. Vessel thickness is uniform throughout the scan. They form a regular spider web.
2. *Deep plexus:* It consists of a close-knit vascular texture, whose orderly distribution around the avascular foveal zone constitutes many fine horizontal and radial interconnections. From each lower extremity, vertical anastomoses arise widening out in the form of horizontal vascular fans that are interlinked to form a complex pattern. There are small vertical anastomoses between the superficial and deeper vessels.

CHAPTER

6

Analysis and Synthesis: Analysis and Interpretation of a Pathological Optical Coherence Tomography-Angiography

Bruno Lumbroso, Marco Rispoli

INTRODUCTION

We develop in this chapter the principles stated in the new terminology section. As in any examination, the study of angiographic images must be carried out following a logical method. Taking this basic principle into account, the reading of an Angio-OCT must be made into two steps:
1. *Analytic steps:* In analyzing an Angio-OCT, the following features must be examined in sequence:
   Localizing the scan depth:
   - Reflectance
   - Flow
   - Morphology and architecture
   - Texture

   Followed by:
2. *Synthesis.*

ANALYTIC STEPS

Localizing the Scan Depth

In studying an Angio-OCT, establishing the depth of the scan in retina and choroid is an essential part of the analysis.

The depth is indicated by the offset. It is negative if it is above the preset segmentation level, positive if it is below the preset segmentation level. The internal limiting membrane (ILM), or Bruch's membrane (BM) that generally corresponds to the retinal pigment epithelium (RPE) ref section (RPEref), is taken as basic profile. Actually also the outer plexiform layer could be used in cases where retina does not present major alterations. However, when retina is deformed, scans taken at plexiform level are not reliable.

The segmentation levels are generally the ILM, the inner plexiform layer (IPL) and the ideal pigmented epithelium (RPEref), or better still BM.

Localizing Scan Depth is not Easy as it Could Seem!

While getting exact level is easy for the healthy eye, in clinical practice the IPL is most frequently deformed, thickened, or unrecognizable. The reference points are usually the ILM, the ideal pigmented epithelium profile (RPEref), or BM.

Interpretations of en-face OCT and Angio-OCT images at a given level are difficult and require a careful, comprehensive and global examination. Retinal pathology can cause displacement of retinal layers from their normal levels. In addition, image processing software errors can result in incorrect identification of reference positions, so sometimes it is necessary to proceed by manual segmentation in order to place correct boundaries.

Correct interpretation may require examination of en face images at multiple depth levels and cross-sectional images through the region of pathology in order to confirm clinical interpretation of an en face image at a specific level.

Reflectance

Reflectance is the contrast for structural OCT. In structural OCT, structure density is shown by variation in reflectance. Structural OCT is sensitive only to the intensity of backscattering and can provide information about blood flow. As a result of this limitation, structural OCT cannot differentiate between vascular tissue and the surrounding tissue.

Table 6.1 Dense hyperreflectance deposits

- Exudates
- Accumulation of pigment
- Thrombotic microaneurysms
- Retinal hemorrhages

Anomalous low reflectance due to masking, screen effect: Anomalous elements may block or decrease vision in Angio-OCT; the boundaries may be sharp (hemorrhages at various levels of the retina, pigment, foreign bodies, etc.) (Table 6.1). More rarely they are indistinct (opacities or hemorrhages of the vitreous) with unvaried size and intensity until the later stages. The screen effect observed in Angio-OCT is generally less evident and less marked than it is in fluorangiography and in cross-sectional OCT (Table 6.1).

Flow, Decorrelation Signal, Blood Cell Motion Signal

The flow is another new and fundamental concept in Angio-OCT that is neither present in classical OCT nor in fluorangiography. Flow is the denomination for contrast in Angio-OCT and is not related to reflectance as seen in structural OCT. Other denominations are decorrelation signal, blood cell motion signal. In Angio-OCT, flow, blood cell motion signal or decorrelation signal refer to the whiteness or albedo that may show some variations in dependence of position of the vessel that may be transverse or perpendicular and blood flow.

Optical coherence tomography-angiography is generated by motion contrast. Blood cell motion is detected using a decorrelation signal from either intensity or phase. Perhaps, it would be better to use the term "decorrelation signal". One could also use the term "Angio-OCT signal".

David Huang and Yali Jia developed the split-spectrum amplitude-decorrelation angiography (SSADA) algorithm to improve flow detection. This technique makes it possible to obtain an Angio-OCT in a few seconds by using the new generation commercial OCT instrument or a prototype.

The SSADA algorithm allows a distinction between blood flow and static tissue. By calculating the decorrelation between consecutive B-scan signals, a contrast is created between the static tissue and the nonstatic tissue that allows us to visualize the blood flow. Flow is therefore highlighted by using the SSADA algorithm.

Movement artifacts are eliminated by using orthogonal recordings and by associating four scans. The flow elements are differentiated from structural data and can be combined into angiograms of several colors in the en-face and B-scan sections. We thus obtain the angiogram of the new vessels, the areas of the new vessels and the flow index of the new vessels.

Structural OCT is sensitive only to the intensity of backscattering and can provide information about blood flow. As a result of this limitation, structural OCT cannot differentiate between vascular tissue and the surrounding tissue.

Optical coherence tomography-angiography examines retinal circulation based on blood flow that may be fast or slow. Its direction is transversal or vertical. The greater the decorrelation, or greater the differences between the consecutive scans as studied by the SSADA, the greater the flow signal. If the flow is too strong, the color becomes black. Measurements in the neovascular area and the flow index give us new information about the neovascular activity and could probably indicate whether there is response to treatment or there may be relapses.

The greater the decorrelation (namely the change in consecutive scans), the greater the vascular signal. There are maximum and minimum values above and below which the color is black (the flow is too strong) or white (as in the exudates).

- *Vascular decorrelation signal:* It is the outcome of an algorithm that represents the retinal circulation. Flow and vessel size do not significantly influence *blood cell motion signal.*
- *Nonvascular decorrelation signal:* There are conditions where the SSADA shows nonvascular reflecting structures that are present in the corresponding en-face imaging as, for instance, hard exudates, pigment accumulation, thrombosed aneurysms and retinal hemorrhages. Further research is needed to understand this effect. Decorrelation signals can be generated by structures, which are not moving blood. The decorrelation signals seem to be generated by pathology that has very fine internal structure, as the examples described.

This might be because very small eye motion, or OCT scanning changes, causes these small structures to generate a decorrelation signal. Larger structures must move much more in order to generate a decorrelation signal.

Flow Velocity

The decorrelation signal has a *limited dynamic range* response to flow. Therefore, one must be careful interpreting the decorrelation signal as a flow velocity. The slowest flow (*sensitivity limit*) that can be detected is determined by the time between repeated B-scans. If the time between B-scans is increased, then the decorrelation signal is more sensitive to slower flows. However, longer time between B-scans means the decorrelation signal is also more sensitive to eye motion and has more noise.

The decorrelation signal cannot differentiate flows which are faster than a certain flow (*saturation limit*), because any flow which is faster will produce the same decorrelation signal. This means that the decorrelation signal is not responsive to changing flow velocity above a *saturation limit*.

For larger vessels, the flows are always fast and generate the same decorrelation signal. The flow is faster the *saturation limit* and decorrelation signal is saturated.

However, for small vessels, such as capillaries or the choriocapillaris, the flow may be too slow and below the *sensitivity limit*. The small vessel with slow flow may not generate a decorrelation signal and will be invisible on the OCT angiogram, even though the vessel is present.

Therefore, if vascular features are not visible on OCT angiograms it may mean that there is dropout or atrophy of the vasculature or that there is reduced/impaired flow.

We should not use the term "reflectivity" as it is usually employed to describe the reflection of light from mirrors or smooth surfaces, like glass. Therefore, the term "reflectivity" may be confusing.

Intensity of decorrelation signal: Flow velocity can change the decorrelation signal. It is also important to note that the decorrelation signal has a limited dynamic range response to flow (Table 6.2).

Morphology and Architecture of the Network

Vessels size and course: Important parameters are the size and course of the veins.

In normal retina, we have:
- *The superficial plexus:* The vessels form multiple linear structures having a centripetal distribution converging towards the fovea while they originate from the upper and lower large vascular arches. Secondary vessels arise from the main vessels, forming a web. The thickness of the vessels is uniform throughout the entire scan. The web is regular without any sudden change in direction and without the formation of vascular rings or loops. The vessels present the same blood signal throughout the scan. Around the avascular area the capillaries form continuous perifoveal arcades with fairly regular meshes.
- *The deep plexus:* It consists of a close-knit pattern of vessels whose orderly distribution around the avascular foveal zone presents numerous subtle horizontal and radial interconnections. The vascular network consists of small vertical interconnection anastomoses between the more superficial and deeper vessels of the same deep plexus. From each lower extremity of the vertical anastomoses short horizontal vessels fan out that interlinked and thus form a complex pattern. An important aspect that cannot be detected with fluorangiography is the presence of normal or anomalous connections between the superficial vascular plexus and the deep vascular plexus.

In pathologic conditions, vessel's size may be regular or irregular while the course may be regular, winding, extended, distorted or meandering in various combinations in the different vascular pathologies.

Vascular network: The architecture of the vascular network may be regular or irregular, and it may be dense, sparse, widened, dilated, rarefied, faint, dense and tangled. The meshes may be sharp, distinct or indistinct. Network may be loose, dilated, narrow, rarefied, faint, dense, compact, tangled, sharp or indistinct.

Vessels shape: We must also assess the shape of the vessels that may be regular or irregular and the density of the capillaries that may be sparse or dense. The cross-section of the capillaries shows up their diameter that may be small, large, regular or irregular (Table 6.3).

Morphological anomalies of the vascular network: The formation of venous loops is caused by vitreous traction that raises a venous segment making it rotate on itself, thus forming a curve, loop or circle (Tables 6.4 to 6.7).

Table 6.2 Blood flow and direction of flow

- Blood flow:
  - Rapid
  - Slow
- Direction of flow:
  - Transverse
  - Vertical

Table 6.3 Aspects of the vessels

- Thin
- Wide

Table 6.4 Morphological anomalies of the vascular network
- Loops
- Anastomoses and shunts

Table 6.5 Architecture of the network
- Irregular
- Tangled
- Uneven
- Network
- Meanders: Increased or decreased
- Bifurcations: Wide or Narrow
- Rarefaction, loss of capillaries
- Proliferation

Table 6.6 Network density
- Larger, smaller meshes
- Irregular
- Tangled
- Uneven
- Dense
- Rarefied

Table 6.7 Vascular network
- Regular
- Irregular
- Widened
- Dilated
- Narrow
- Rarefied
- Faint
- Dense
- Compact
- Tangled
- Sharp or Indistinct

Network density and shape features: The *density* of the vascular network of any Angio-OCT retinal and choroidal layer needs to be studied separately, as do the sizes of the vessels that may be large or small, irregular, with or without dilatations. There may be:
- Meanders: Increased or decreased
- Bifurcations: Wide or narrow
- Rarefaction
- Capillary loss
- Capillary dropout
- Vascular proliferation
- Also vessel size must be examined (large or thin) as well as capillary density.

Table 6.8 Microocclusions, capillary dropout
- Diabetic retinopathy
- Vein occlusion
- Vasculitis
- Eales' disorder
- Thalassemia
- Sickle-cell anemia
- Other blood disorders

Rarefaction of the Vascular Network in Angio-OCT (Table 6.8)

In *diabetic retinopathy*, areas of ischemia may be observed at the posterior pole or in the midperiphery.

In *other ischemic retinopathies*, Eales' disease, sickle-cell anemia, ischemic areas located are mainly in the periphery and hence difficult to observe using Angio-OCT.

Artery occlusions: They include rarefaction or disappearance of the vascular network; extremely marked thinning or disappearance of the vessels lumen.

Edema raises and deforms the already sparse capillary network causing dilatation of the vascular meshes.

Anomalies of the Perifoveal Capillary Network in Angio-OCT (Tables 6.9 to 6.15)

Alterations of the perimacular vascular network: In diabetic patients, the finding of a particularly evident and sharp macular capillary network points out an incipient retinopathy. Indeed, changes occur in the macular capillary network already at the extreme onset of diabetic retinopathy. This feature is due to the increase in the size of some capillaries, while others are closed and thus form a looser network with larger and sparse meshes.

Table 6.9 Vascular network
- Meanders: Increased or decreased
- Bifurcations: Wide or narrow
- Rarefaction, capillary dropout
- Proliferation

Table 6.10 Size of vessels
- Large or small
- Irregular
- Dilatations
- *Vascular dilatations:* Irregular, microaneurysms, macroaneurysms

Table 6.11 Shape of vessels
- Regular
- Irregular

Table 6.12 Density of capillaries
- Sparse
- Dense

Table 6.13 Section of capillaries, diameter
- Small
- Large
- Regular
- Irregular

Table 6.14 Vascular network
- Dense
- Sparse
- Wide
- Dilated
- Rarefied
- Dense
- Tangled

Table 6.15 Meshes of the capillary network
- Normal, regular
- Sharp meshes
- Distinct or indistinct meshes
- Network more visible than normal (diabetic retinopathy)
- Dilated meshes
- Widened meshes (edema)
- Thin meshes (ischemia)
- Irregular meshes, shunts, loops (ischemia)
- When meshes disappear, grayish texture (ischemia)

There is *an increase of the avascular foveal area size* that is normally 500 μm. This is an early clinical symptom and it is still reversible. As the retinopathy evolves, in the absence of edema that alters its features, the macular capillary network will become increasingly evident and more marked alterations start to appear: slight congestion of the capillaries and some dilation. The presence of small ischemic areas at the posterior pole leads to the occlusion of the smaller vascular branches; the network becomes looser at first and later the small areas of ischemia grow larger and merge; *interruptions in the macular capillary arcades* appear that will progressively extend.

Capillary lesions: Capillary lesions are often observed in the nonperfused retinal areas; the arterial capillaries are thin. From a normal arteriole branch there arises a vascular stump that is immediately interrupted at the edges of a hypoxic zone. There are also venous capillary dilatations, loops of capillaries, microaneurysms, arterial-venous anastomoses (shunts) and roundish, deep hemorrhages.

Venous alterations: Retinal veins may present anomalies in their dimensions and shape, they may be dilated for long stretches or they may present segment ectasia or pearl necklace, sausage-like aspects. With Angio-OCT the blood column can be studied.

In the ischemic areas, arteriovenous anastomoses are observed, formed by segments of preexisting capillaries, dilated in such a way as to create short circuits between veins and arteries while other capillaries are closed and not visible at angiography. Duplications are formed by collateral veins, running parallel to a venous trunk that becomes progressively larger, until they replace the occluded. Venous alterations are almost always associated with areas of ischemia.

Texture

Texture *is a new concept in description of Angio-OCT that is not used in classical structural OCT or in fluorangiography.* Texture is the feature made by the threads of a fabric. In a pavement, it is the aspect produced by its components; it could be fine or coarse and loose. This concept is used extensively in modern radiological techniques. The texture needs to be evaluated layer by layer and is related to the presence or absence of capillaries.

Texture may be coarse, granulated, loose, fine, faint, subtle, speckled and grayish. Texture is a very interesting concept.

However, we should be careful how to define it, because it can change with image averaging. For example, fine texture in retinal layers changes on some other commercial device images depending on the averaging. Coarse texture is less sensitive to averaging effects. Since different instruments in the future may have different amounts of averaging, it would be good to define texture so that it works with future instruments (i.e. use features that are not so sensitive to image averaging).

The texture should be assessed layer by layer and is related to the presence or absence of the vascular plexus that may be altered or absent.

Table 6.16 Texture
• Loose • Coarse • Large texture • Granulated • Fine • Faint • Speckled • Subtle • Thin • Grayish

In Angio-OCT, the differences in texture are evident in the case of ischemia or edema.

- *Ischemia:* It shows a total disappearance of the capillaries, the area has a grayish aspect and the texture is more or less fine or finely granulated. The density of the collateral branches depends on the extent of the ischemia. Connections (shunts) between the retinal vascular superficial and deep plexuses are almost invisible in normal conditions. They are, instead, extremely frequent and evident in the vascular pathology of the retina. The density of collateral branches depends on the ischemia. In ischemic areas, the margins are sharp. Hypoperfusion or vascular filling defect is characterized in Angio-OCT by a grayish area, with a finely granulated texture that is different from a normal texture, and this is evident by way of contrast with the normal texture of the background.
- *Microocclusions of the capillaries:* Diabetic retinopathy, venous thrombosis, vasculitis, Eales' disease, thalassemia, sickle-cell anemia and other blood disorders. In diabetic retinopathy, the areas of hypoperfusion can be observed at the posterior pole or in midperiphery. In the other ischemic retinopathies, like Eales' disorder or sickle-cell anemia, the areas of hypoperfusion are mainly in the periphery.
- *Edema:* Edema texture appears as a grayish area related to the widening of the vascular network meshes and rarefaction of the finer capillaries. The margins of the edematous areas are sharp. Edematous area is less dark than normal retina (Table 6.16).

ANALYSIS OF NEOVASCULAR MEMBRANES

The extreme neovascular membranes heterogeneity as regards their shape, location and flow constitute a separate chapter in Angio-OCT analysis.

Localizing Depth

Depth (offset) is the first element to be taken into account because it is necessary to localize exactly the position of the neovascular membrane in regards to retina layer (preretinal, intraretinal and subretina), if macular, or peripapillary. Location with respect to pigment epithelium and BM is very important. Occult (type 1) membranes are placed below RPE; classical (type 2) membranes are placed above RPE. Localization has important influence on diagnosis and treatment decision making.

Profile

Next we need to assess the *profile* of the membrane in order to choose the best segmentation. The correct segmentation of a neovascular membrane does not always coincide with the correct segmentation of the surrounding retina.

Thickness

The *scan thickness* is crucial for assessing the neovascular membrane. Indeed, the new vessels may have a rather variable thickness in that they do not develop regularly but in an uneven shape.

The *section thickness* is extremely important also with regard to the average size of the new vessels. If the new vessels are rather thin, a segmentation too thick would produce a confused image of the flows at several levels.

It is of the utmost importance in the segmentation of a neovascular network to verify that we see truly a vascular network and not a "mirror effect" at the level of the RPE. At times there is risk that the neovascular network may be confused with some atypical branches of the superficial plexus projected onto the RPE. In order to avoid this inconvenience it is sufficient to perform the analysis dynamically closer to the surface and then in depth to check the disappearance (if this is the case), of flows that were a "mirror effect".

Flow

Age-related macular degeneration: In this disorder, the new vessels are found mainly just beneath the pigmented epithelium above BM (type 1) and in a smaller percentage above RPE (type 2). In Angio-OCT we speak of evidence of flows and only indirectly do we talk about actual vessels.

Table 6.17 New vessels: Stage of development
- Tree-like
- Cartwheel
- Capsular
- Avascular fibrotic scar

Table 6.18 New vessels: Morphology
- Bicycle wheel
- Cartwheel
- Octopus like
- Tree-like
- Glomerular
- Whirling
- Fragmented

In naive patients, the flows appear to be branching with a main feeder trunk. They may be rounded, tree-like, octopus-like or cart-wheel shaped (Tables 6.17 and 6.18). The peripheral anastomoses may be evident and constitute an important parameter to be assessed in the follow up.

These flows have a *clear, evident* aspect and it is possible to follow precisely the course of the main and secondary branches. The flow in neovascular membranes seems to be more evident with respect to the normal vessels adjacent flows.

After treatment with anti-vascular endothelial growth factor (VEGF), the network will appear to be of a smaller size, with less anastomoses but the main changes after treatment is the network *fragmentation*. In a fragmented flow it is difficult to follow the vascular branches course. Secondary branches disappear immediately after the intravitreal level and then reappear after a few weeks.

SYNTHESIS

Within the logic strategy of Angio-OCT interpretation, the synthetic study follows the analytical one and constitutes a most important step.

An isolated study of OCT angiograms cannot yield precise and global information. The only way to really understand and evaluate Angio-OCT requires a global, omni comprehensive study of the subject's pathology.

The analytical breakdown of structural OCT scans includes a study of the morphology (i.e. morphological changes, anomalous structures), and of the variations in reflectivity (i.e. hyperreflectivity, hyporeflectivity and the shadow effect), structure, segmentation, texture, tridimensional, "en face" and the quantitative study (thickness, map, volume).

It now with the Angio-OCT includes also the *flow, morphology, architecture and texture.*

The subsequent synthesis and deduction step combines the elements of the OCT analysis with the other data supplied by the medical history, the clinical examination, autofluorescence, structural OCT.

Putting together all these data make possible the deduction and the diagnosis.

CHAPTER 7.1

Clinical Applications: Aspects of OCT SSADA Angiography in Eye Disorders

Bruno Lumbroso, Marco Rispoli, Maria Cristina Savastano,
Adil El Maftouhi, Leonardo Mastropasqua, Luca Di Antonio, Giovanni Staurenghi

INTRODUCTION

Optical coherence tomography-angiography (Angio-OCT) clinical applications are many, and their number is increasing. Three main categories presently under study are: (1) retinopathies, (2) vascular acquired or congenital malformations and (3) choroidal neovascularization (CNV). We will study Angio-OCT disorders separately in the Inner Retina and in the Outer Retina.

PART I — INNER RETINA

AGE-RELATED RETINAL ANOMALIES

Aged persons show some alterations at deep vascular plexus level (Figs 7.1.1A and B). Angio-OCT shows that the main superficial retinal vessels are normal. At the level of the deep plexus, the capillaries are rarefied. Vessels lose part of their collateral branches. The capillaries have the aspect of irregular fans.

RETINAL ANOMALIES AND COAT'S DISEASE

Coat's disease and Leber's forms of telangiectasia are two aspects of the same syndrome. They present telangiectasia and aneurysmal vasodilation. In the later stages of the disorder, exudates and exudations appear.

Optical coherence tomography-angiography shows that the main superficial retinal vessels are typically tangled.

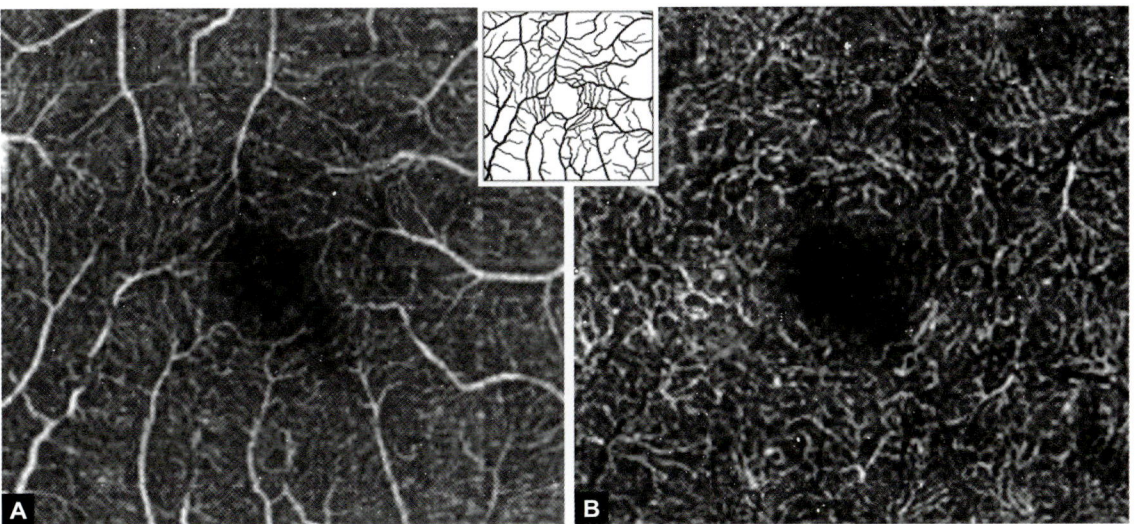

Figs 7.1.1A and B Superficial (A) and deep vessels (B): in this 65-year-old individual the deep vessels appear to be more sparse and slightly irregular

At the level of the superficial plexus, the vessels lose most of their collateral branches and present many harmonious-shaped loops. The capillaries are rarefied and anomalies in vessel size, vasodilation and macroaneurysms are evident (Figs 7.1.2 and 7.1.3).

At the level of the deep plexus, the capillaries are more rarefied. Size alterations, flow alterations and morphology anomalies of the plexus are evident. The capillary fans are very irregular. In the context of the deep plexus we can see vascular flow anomalies (vasodilation) that are more evident in depth (Figs 7.1.4 to 7.1.7).

RETINAL ANOMALIES IN ANGIOMATOSIS

Other forms of angiomatosis show telangiectasia and aneurysmal vasodilation. Later in evolution, exudates and exudations will appear.

Optical coherence tomography-angiography shows that the main superficial retinal vessels lose most of their collateral branches and present many loops. The capillaries are rarefied, and anomalies in vessel size, vasodilation and macroaneurysms

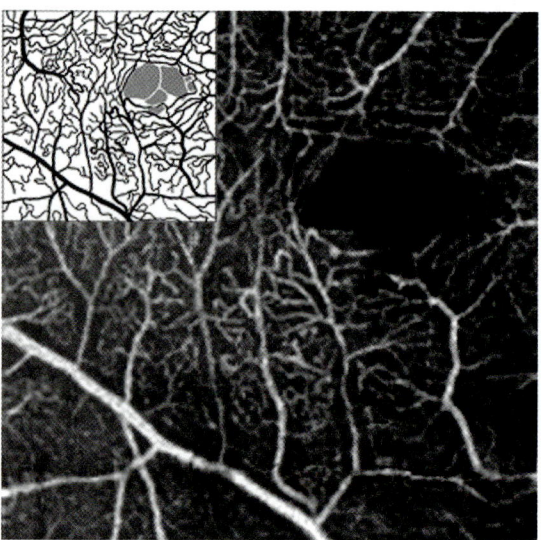

Fig. 7.1.2 Coat's disease, superficial plexus. The optical coherence tomography-angiography (Angio-OCT) of the superficial plexus shows pronounced vascular tortuosity. The vessels have lost most of their collateral branches and present many loops, and regular harmonious course. The capillaries are rarefied with anomalies in size, vasodilation and macroaneurysms

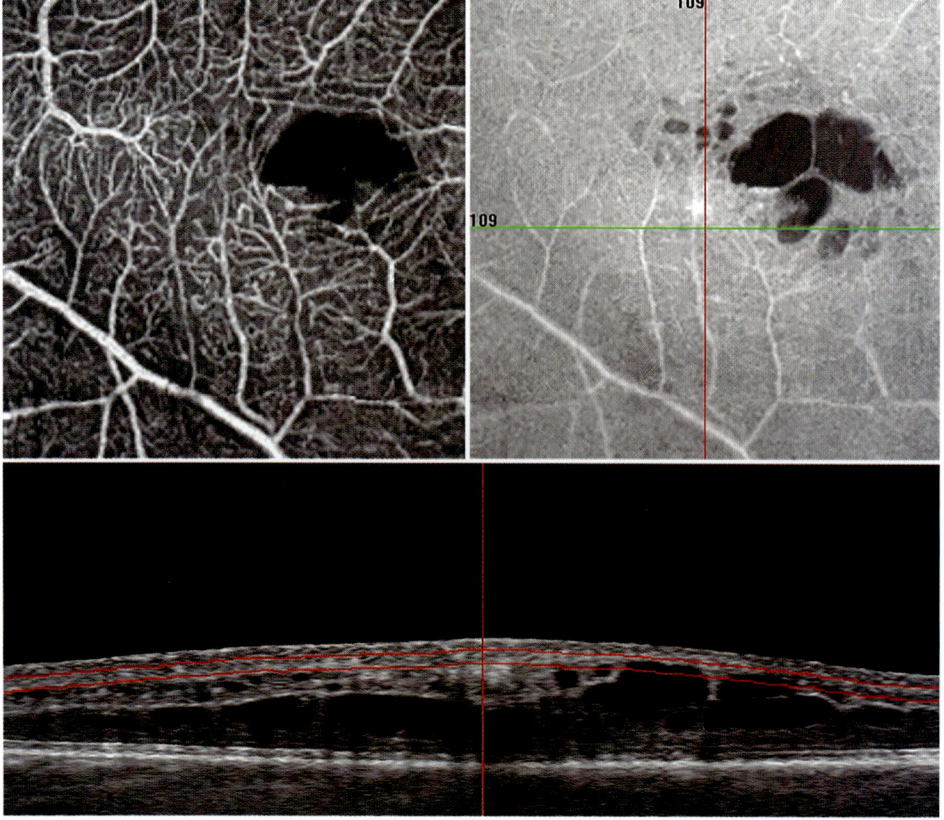

Fig. 7.1.3 Coat's disease: Small cystoid edema cells can be observed at the fovea. Capillaries present anomalies in density, size, vasodilation and macroaneurysms

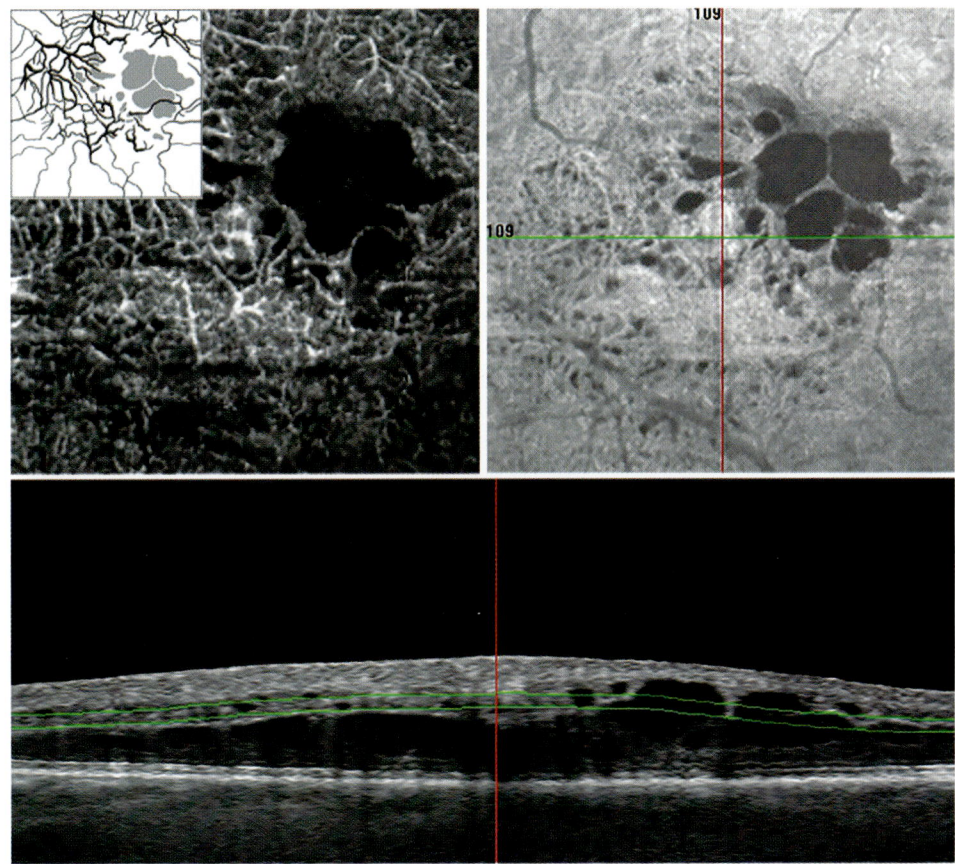

Fig. 7.1.4 Coat's disease, deep plexus. Capillaries are rarefied with evident alterations in size and morphology. The fan-like pattern is very irregular, and flow shows vascular anomalies and roundish vasodilation

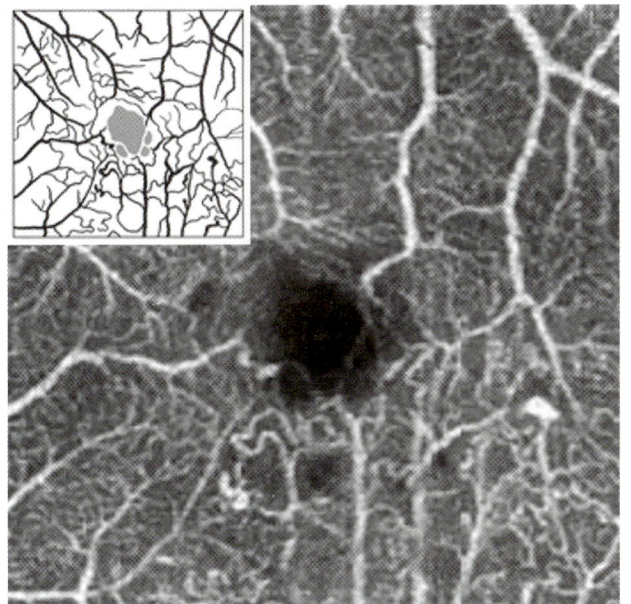

Fig. 7.1.5 Coat's disease, superficial plexus with rarefied capillaries, numerous loops and vasodilation with regular harmonious vessels

are evident. At deep plexus level, size alterations, flow alterations and morphology anomalies of the plexus are evident. The capillary fans are irregular (Figs 7.1.8 to 7.1.11).

RETINAL SUPERFICIAL ANOMALIES IN MACULAR PUCKER

Pucker retinal folds and retraction cause course anomalies at the level of the superficial plexus (Fig. 7.1.12). The vessels lose most of their normal spidernet aspect and mostly follow the folds course.

MACROANEURYSM

Macroaneurysm is seen in Angio-OCT as a rounded cavity located at the deep retinal vascular plexus level. It is surrounded by cystoid edema cells. 15 days after laser treatment, the vascular anomaly disappears, and edema cells decrease. Note the star-shaped hard exudates at this retina level (Figs 7.1.13 and 7.1.14).

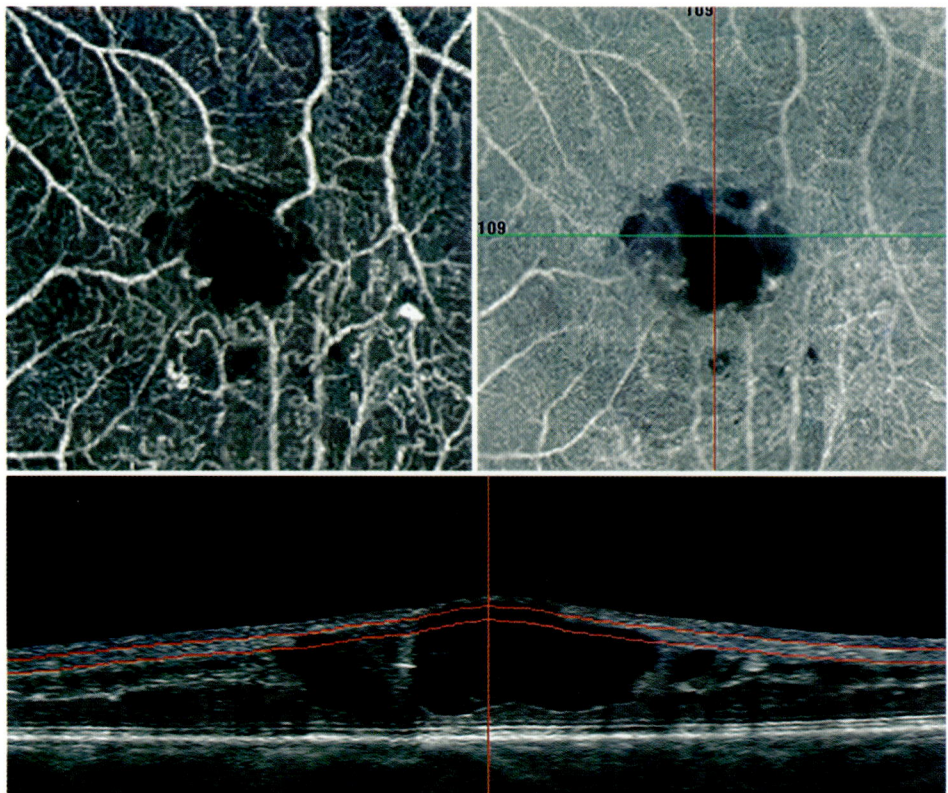

Fig. 7.1.6 Coat's disease, macular edema with very large cystoid cells at the inner and outer macular layers

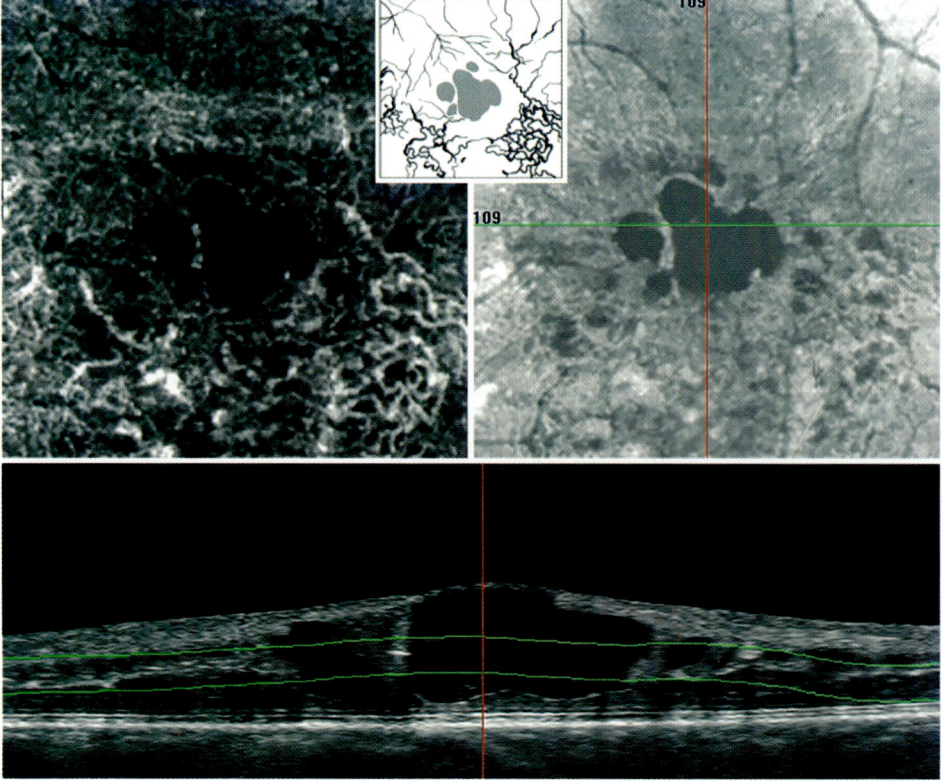

Fig. 7.1.7 Coat's disease, deep plexus, flow anomalies, vasodilation, vascular loops and small cystoid edema cells can be observed

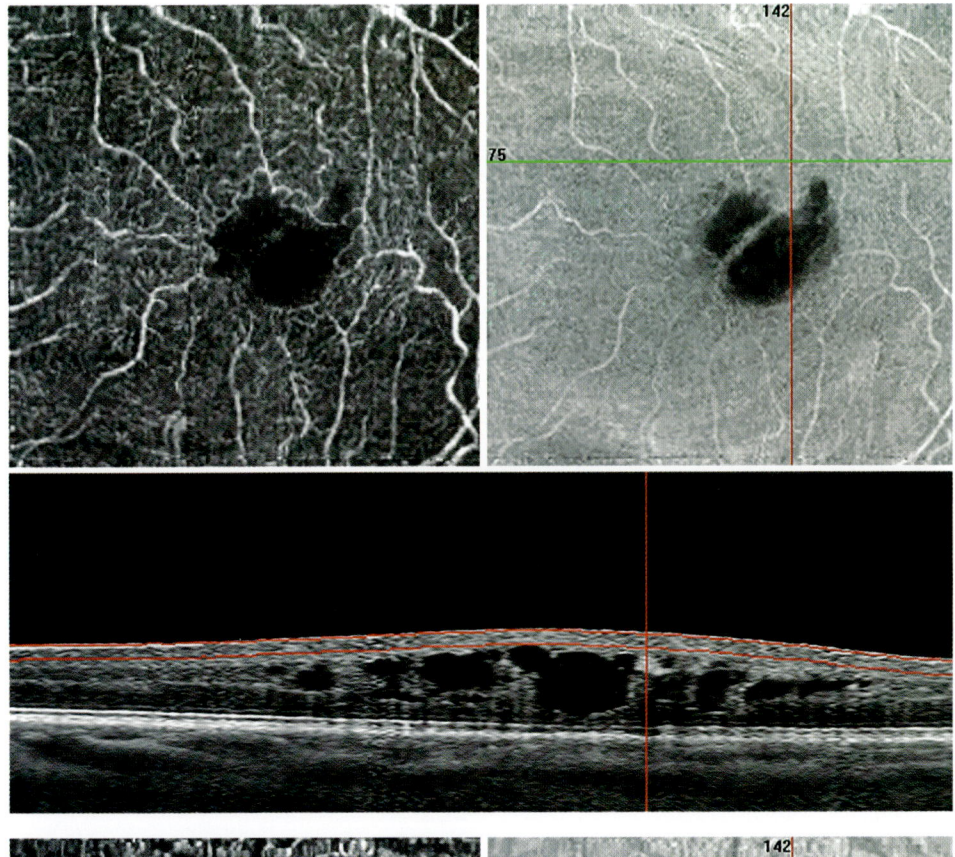

Fig. 7.1.8 Retinal angiomatosis, superficial plexus. The vascular texture is finer than in normal eyes. Vessels course (flow) is irregular, but we do not see the numerous harmonious loops normally present in Coat's disease; presence of cystoid edema

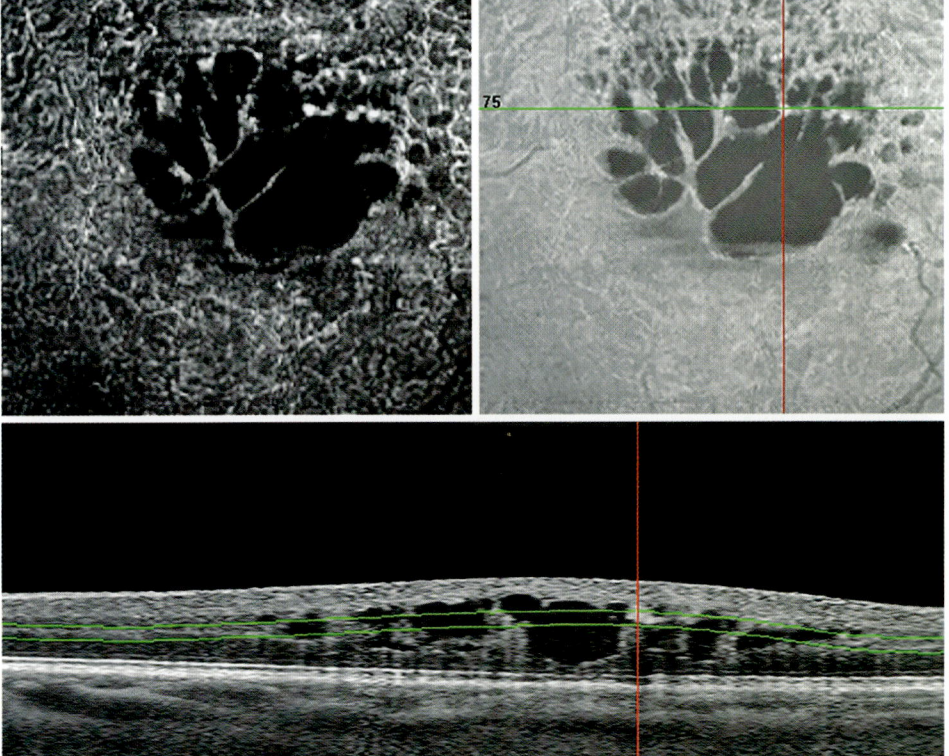

Fig. 7.1.9 Retinal angiomatosis, deep plexus. The normal vascular texture is disrupted by many small cystoid edema cells. Between edema cells, irregularities in the vascular texture, vasodilation and macroaneurysms can be seen

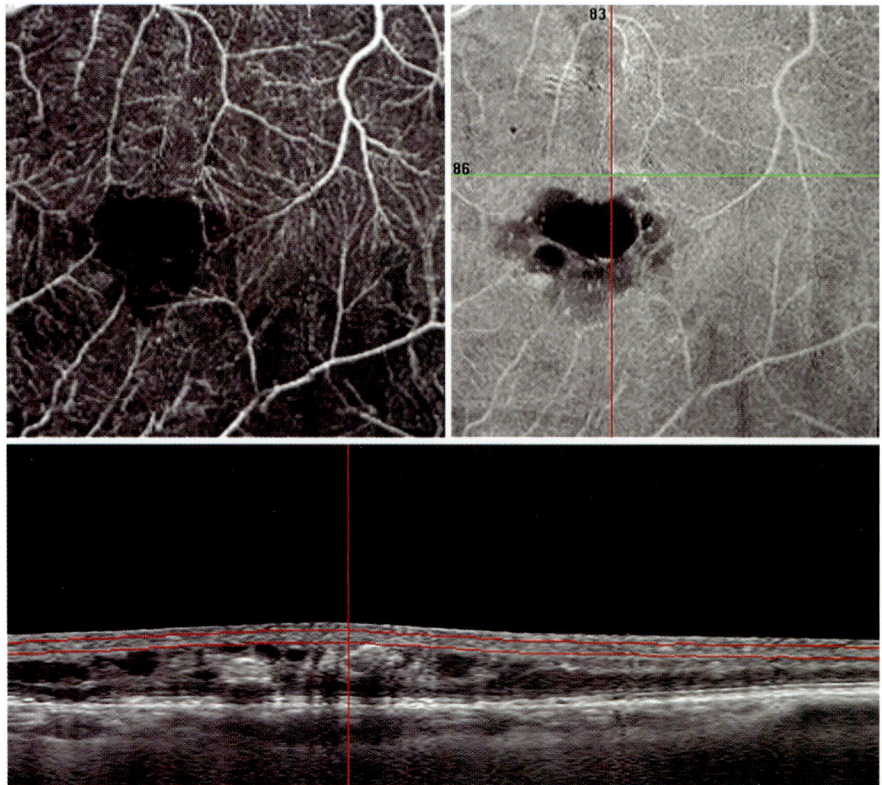

Fig. 7.1.10 Macular telangiectasias. Superficial vascular plexus is almost normal with slight irregularities of the vascular texture

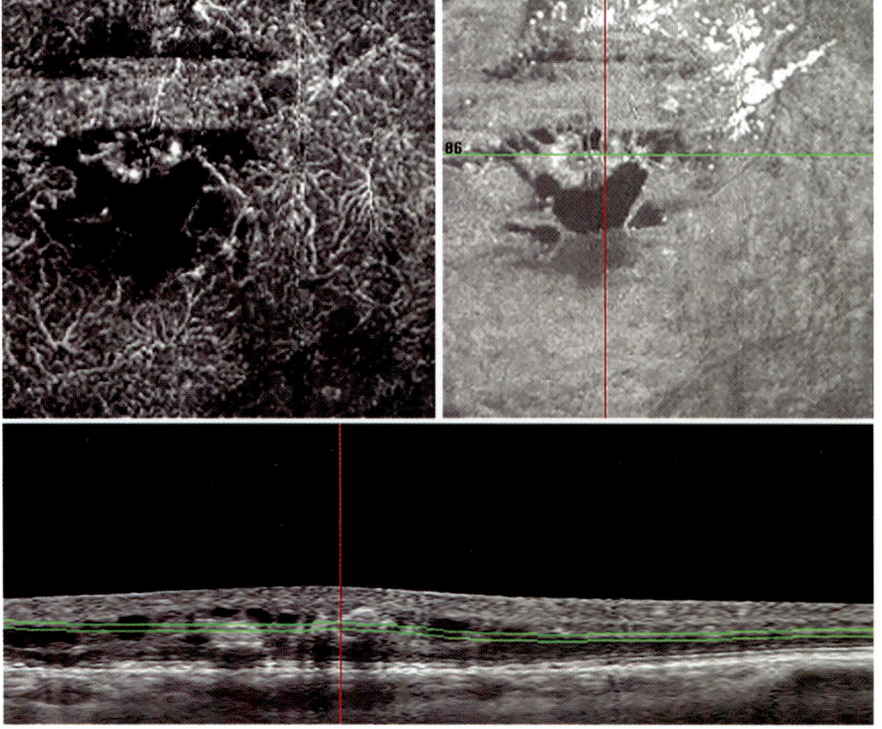

Fig. 7.1.11 Macular telangiectasias, deep vascular plexus; presence of cystoid edema cavities: these cavities have irregular rough walls. Alterations of the deep vascular texture and vessel dilation

Clinical Applications: Aspects of OCT SSADA Angiography in Eye Disorders ■ 29

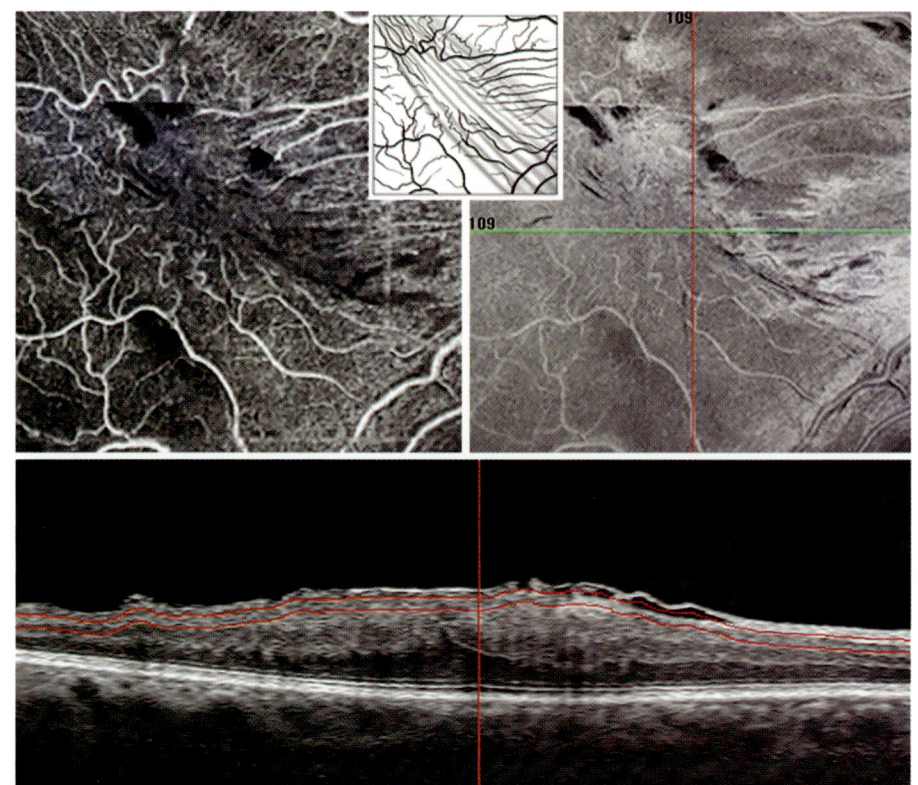

Fig. 7.1.12 Macular pucker. The superficial vascular network of the retina is of normal size, but the course of the macular capillaries is greatly altered by the retinal traction due to parallel retinal folds. Vessels are stretched along the retinal folds, when they are parallel to it, or they present a very irregular wavy course when they cross the folds

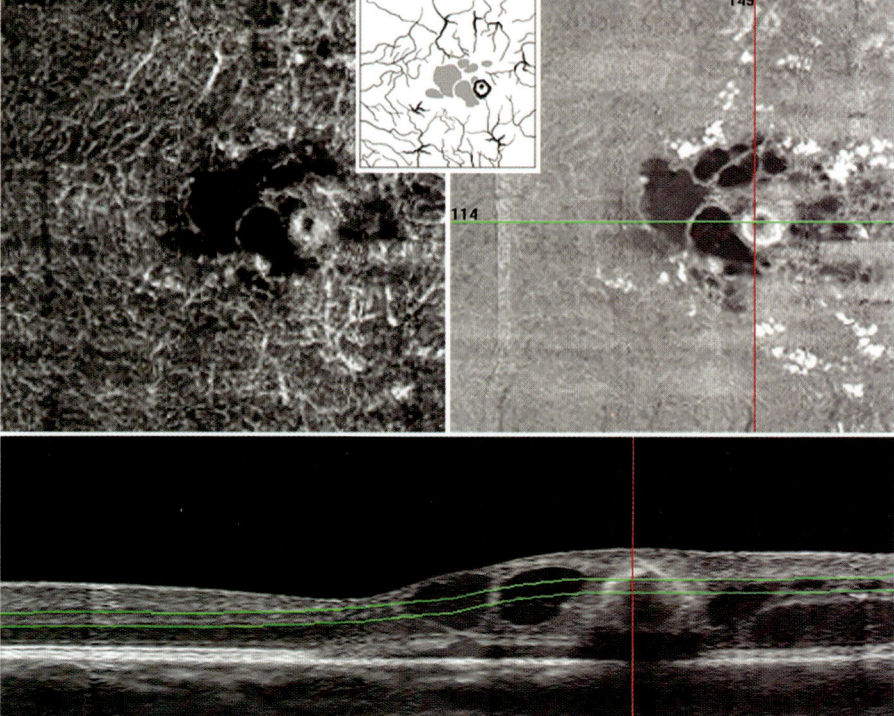

Fig. 7.1.13 Macroaneurysm: It is seen as a rounded formation located in the deep vascular plexus; evident cystoid edema around the vascular lesion

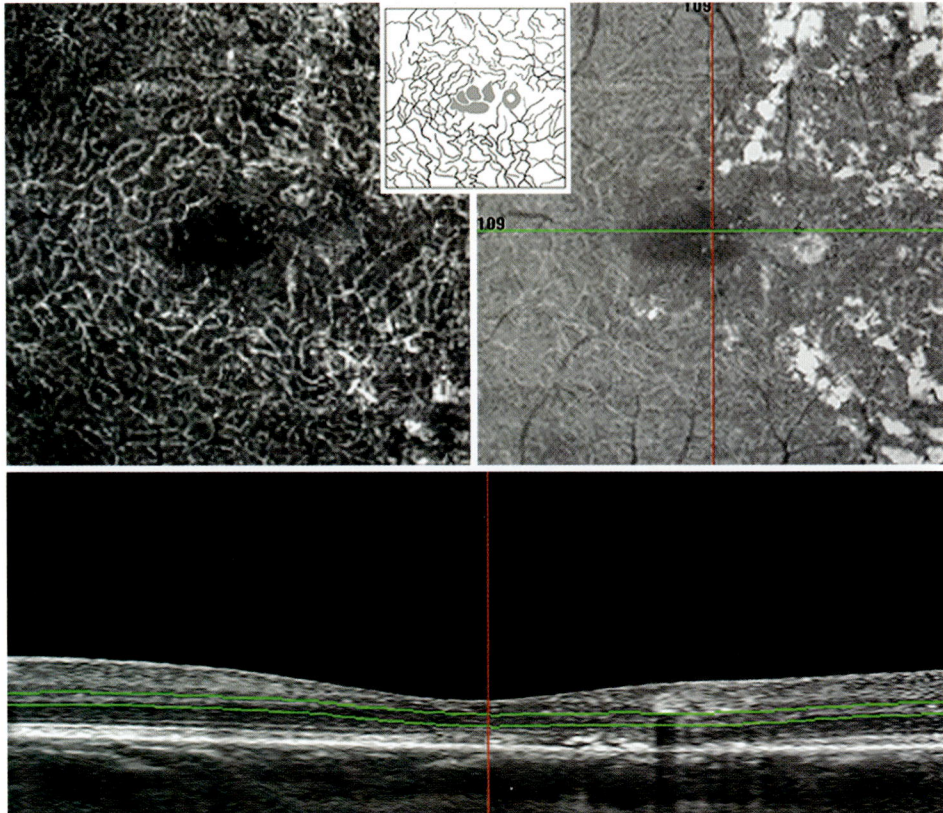

Fig. 7.1.14 Macroaneurysm 15 days after laser treatment and occlusion of the vascular anomaly. The lesion is no more visible on optical coherence tomography-angiography (Angio-OCT). Instead it is still clearly evident on en-face OCT. Cystoid edema has totally regressed, the deep vascular plexus is back to normal; there are still deep exudates

DIABETIC PATIENTS WITHOUT RETINOPATHY

In diabetic patients, even in the absence of retinopathy, Angio-OCT shows the retinal capillaries with greater evidence than in healthy individuals. The avascular foveal area is larger than in healthy individuals.

A particularly evident and sharp macular capillary network is suggestive of an incipient retinopathy because, even before the onset of diabetic retinopathy, there are changes in the macular capillary network. These changes are due to the increase in the size of some capillaries while others are closed, and thus we see a looser network with larger and more sparse meshes.

There is an increase in the size of the foveal avascular area that normally is about 500 μm large. This is an early sign that appears before the onset of the microaneurysms, and at this stage the condition is still reversible. As retinopathy evolves, the capillary network of the macula becomes increasingly evident and more marked alterations will appear, such as mild congestion of capillaries and some dilation. The presence of small nonperfused areas at the posterior pole will lead to the occlusion of small branches; the network becomes at first more irregular, and later, the small areas of ischemia will grow and then merge with the central enlarged avascular area. The deep vascular plexus shows disruption and irregularities in the deep capillaries, with smaller and more sparse vascular fans (Figs 7.1.15 and 7.1.16).

BRANCH RETINAL VEIN OCCLUSIONS

In eyes affected by branch retinal vein occlusion (BRVO), Angio-OCT highlights the vascular network with evident areas of nonperfusion that correspond to areas of nonperfusion in fluorangiography. These areas, however, look sharper because there is no masking effect due to dye leakage in intermediate and late stages of the analysis. Some capillaries increase in size while many more are closed. We see thus a looser network with larger and more sparse meshes and a fine, grayish texture.

The vascular network is seen more sharply, and the arteriovenous anastomoses and vascular loops are easier to see. In addition we can observe details that do not show in fluorangiography because dye leakage hides them in the intermediate and later stages of the examination.

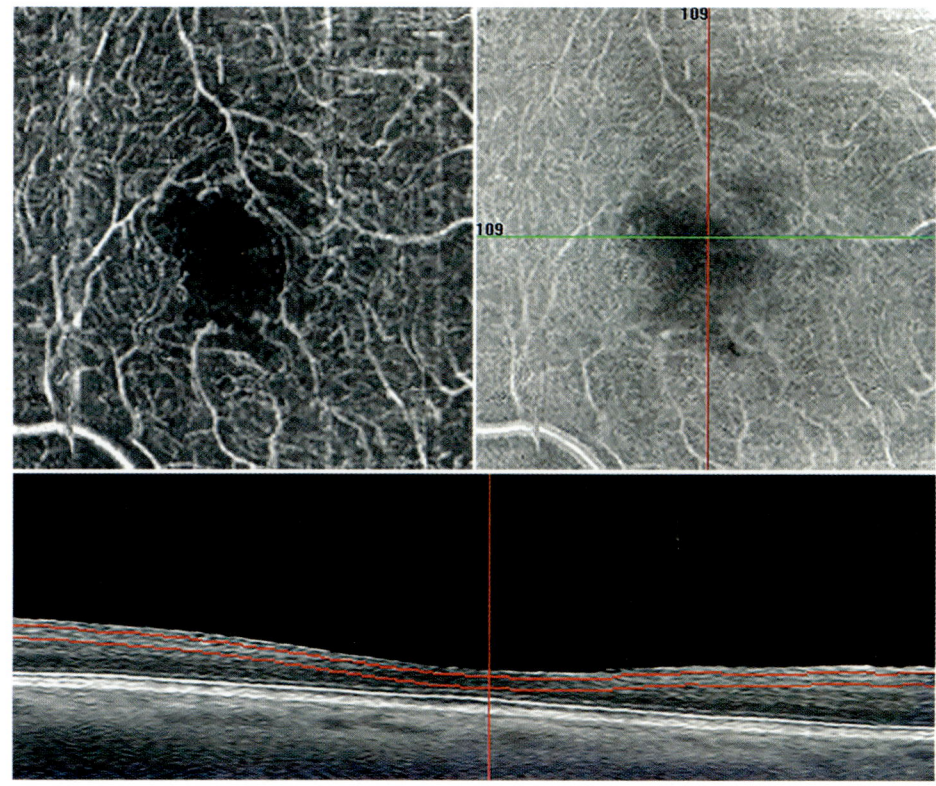

Fig. 7.1.15 Nonproliferative diabetic retinopathy, optical coherence tomography-angiography (Angio-OCT) of the superficial vascular plexus. Marked irregularity in the vascular network with disappearance of the thinner capillaries, enlarged avascular zone with some interruptions in the perifoveal vascular arcade

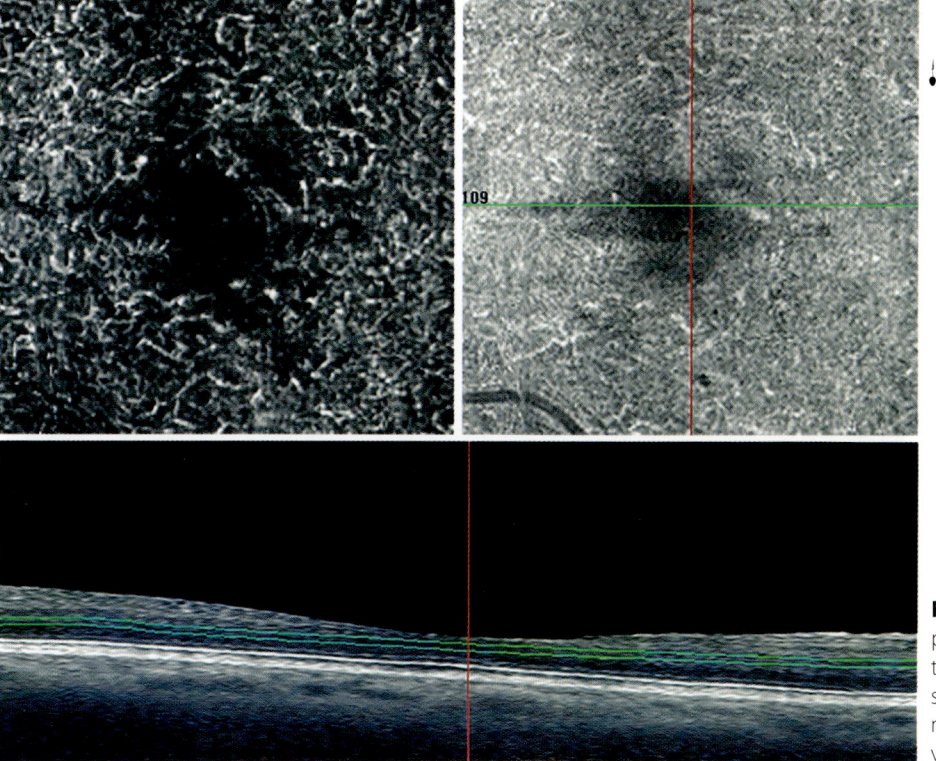

Fig. 7.1.16 Same patient, deep vascular plexus. Disruption and major irregularity in the deep network with uneven capillaries, smaller and more sparse vascular fans, microaneurysms with intraretinal microvascularization

The nonperfused areas appear as zones where the capillaries are sparse and more evident against a gray background. We have to observe the texture that varies from fine to loose. Often the capillaries inside the nonperfusion areas are truncated with abrupt interruptions or there are arteriovenous shunts, or shunts inside the deep vascular network, in the inner nuclear layer.

Retinal edema areas can be easily seen because there is no dye staining. However, at times, we see a widening and distortion of the capillary network meshes and a decrease in the sharpness of the widened capillaries.

When the fluorangiography shows staining of the vascular walls the Angio-OCT instead shows a very thin vessel (that corresponds to the vessel lumen) surrounded by a darker area that corresponds to the thickened vascular wall. Therefore, in this case, there is a sharp visual difference between fluorangiography and Angio-OCT.

Retinal hemorrhages are visible as masked areas, but they are much less evident than in fluorangiography.

In ischemia, the texture of the background may vary from a faint gray to a grayish granulation.

In vein occlusions we see changes in the structure of the superficial plexus, especially in macular ischemia. In these cases indeed, the vascular signal (flow) is not linear but has focal deviations, the wall thickness is not regular but shows focal segmentation and lumen narrowing; the vessels course shows abrupt interruptions with some dilation around the avascular foveal area that appears to be widened with respect to healthy individuals. Vessel flow can be segmented.

The deep plexus varies significantly with considerable differences, especially in the ischemic areas. Capillaries distribution is irregular with various changes in vessel course in nonperfused zone. The wall vessels are thicker in the pathologic area; the vessels course shows multiple shunts along various retinal planes. Texture aspect is different in the ischemic zone (Figs 7.1.17 to 7.1.22).

RECENT OR LONG-LASTING RETINAL ISCHEMIAS

Recent Retinal Ischemias

With Angio-OCT it is possible to highlight the main superficial retinal vessels that characteristically lose most of the collateral branches after the ischemic event. In recent ischemias, this aspect concerns almost exclusively the superficial vascular plexus, with a fair conservation of the morphology, size and vascular signal of the capillaries of the deep plexus.

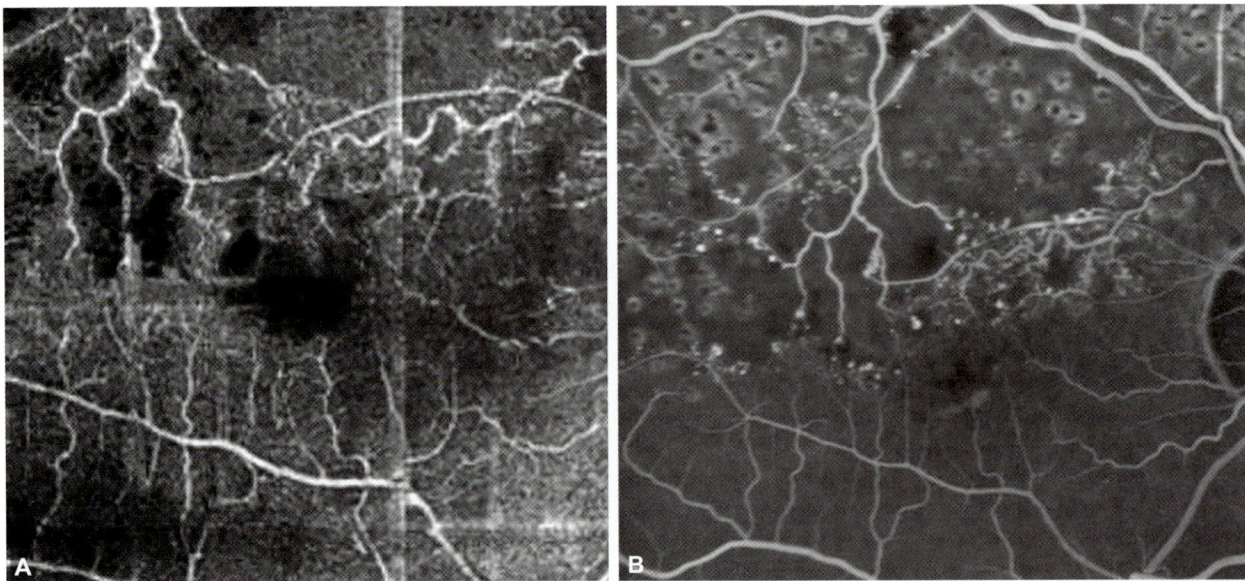

Figs 7.1.17A and B Fluorangiography (B) and Angio-OCT (A) comparison between traditional angiography and optical coherence tomography-angiography (Angio-OCT) in a subject suffered from branch retinal vein occlusion (BRVO). Both techniques clearly show the ischemic areas. Angio-OCT shows evident fading of the surface texture as a grayish background. Retinal vessels are tortuous while edematous areas appear as black areas (very low or no decorrelation)
Courtesy: Giovanni Staurenghi

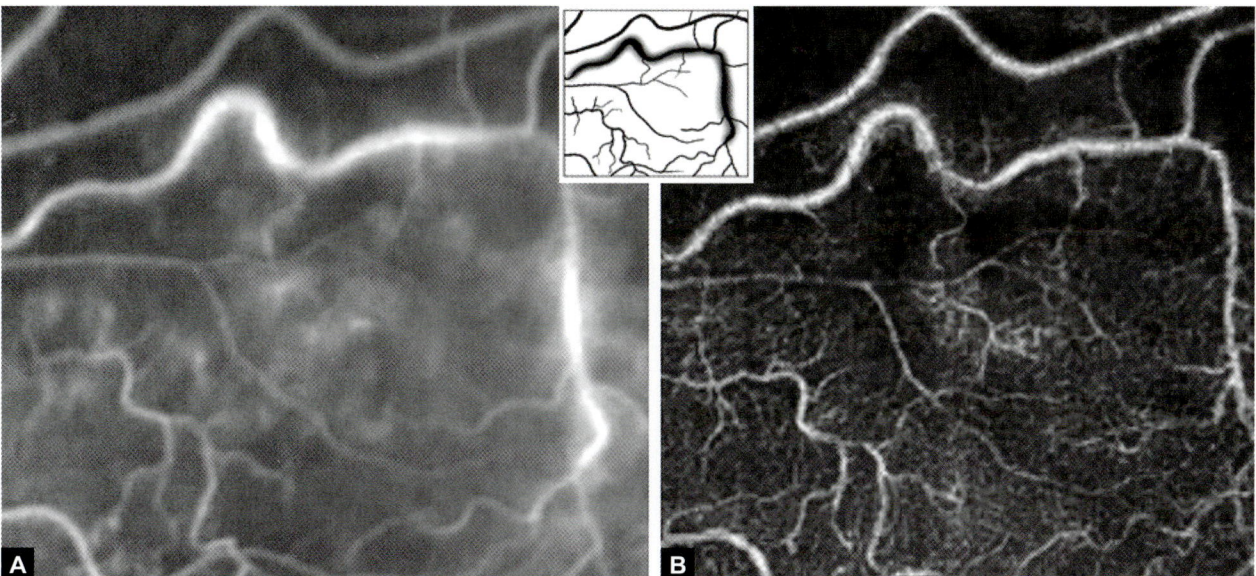

Figs 7.1.18A and B Branch vein occlusion. (A) Posterior pole fluorangiography: Hyperfluorescence of the altered vascular branches and dye leakage around the vessels can be observed. In figure, top is a grayish area of ischemia, in the center we observe diffuse hyperfluorescence areas due to retinal edema. (B) Same area of optical coherence tomography-angiography (Angio-OCT): Vascular flow is evident but vessels diameter is greatly reduced as shown by the blood column (the wall cannot be seen). There is neither leakage nor staining. The upper ischemic area is dark with marked decrease in vascularization and smooth, almost grayish texture. At the posterior pole, in the edematous zone, the capillaries are more numerous albeit irregular and texture is coarse

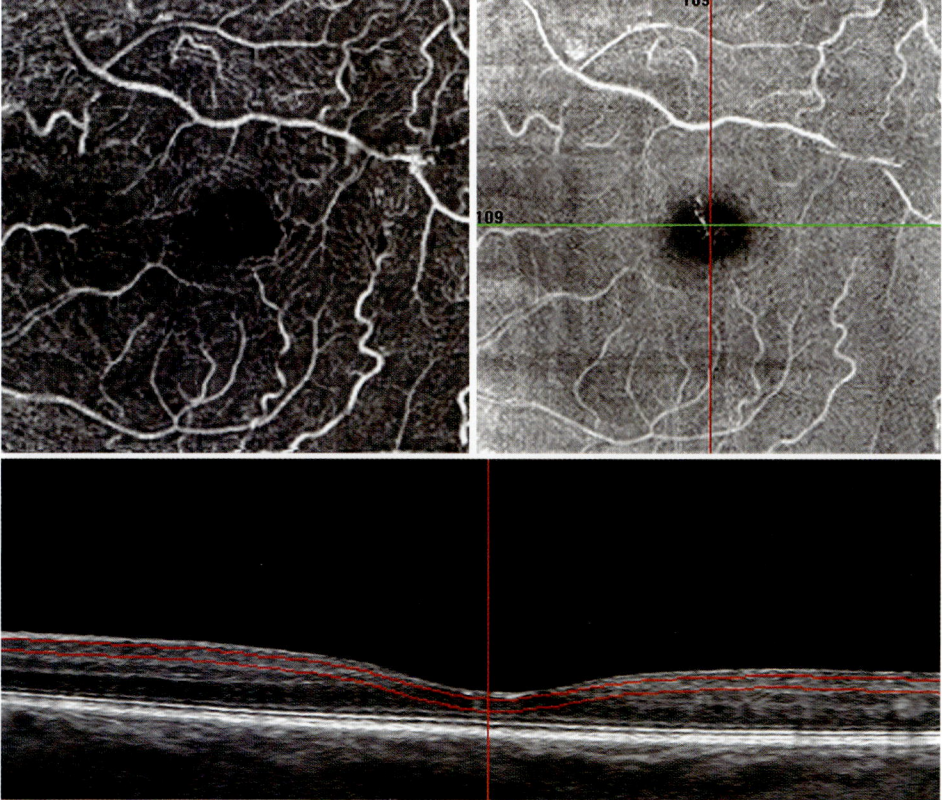

Fig. 7.1.19 Branch vein occlusion. Optical coherence tomography-angiography (Angio-OCT) shows up a finely granular vascular texture at superficial vascular plexus level, with larger avascular foveal area and marked vessels irregularity

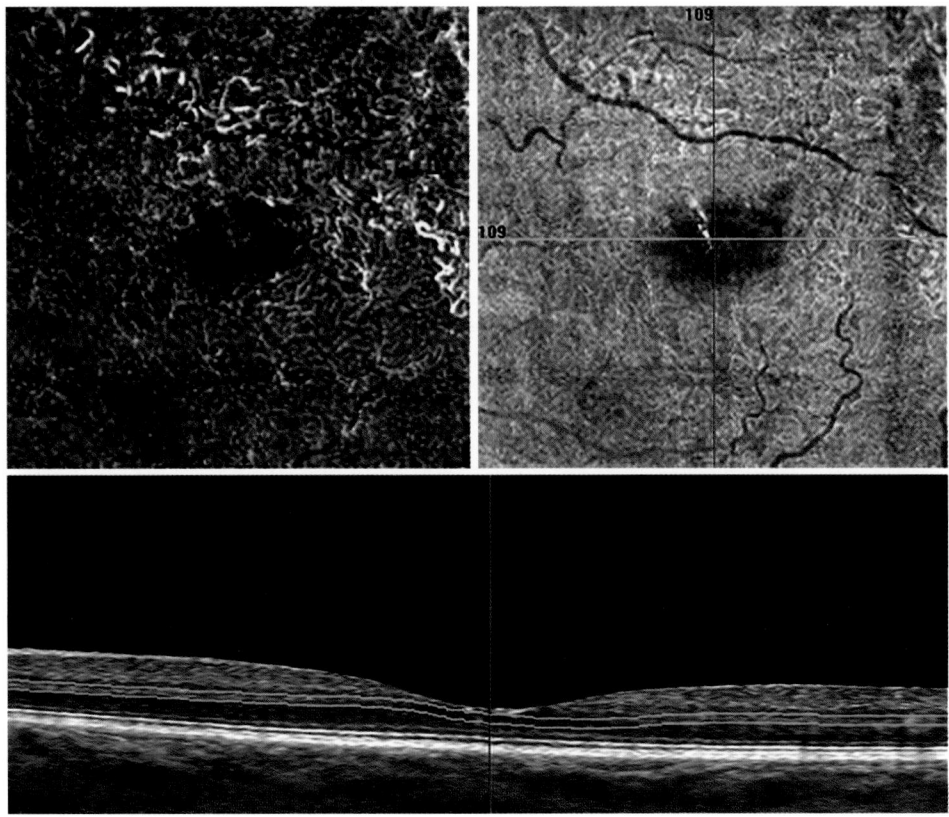

Fig. 7.1.20 Branch vein occlusion, deep vascular plexus of same eye seen in Figure 7.1.19. Marked alteration of the deep texture where capillaries are fragmented with disruption of normal regular fan-like pattern

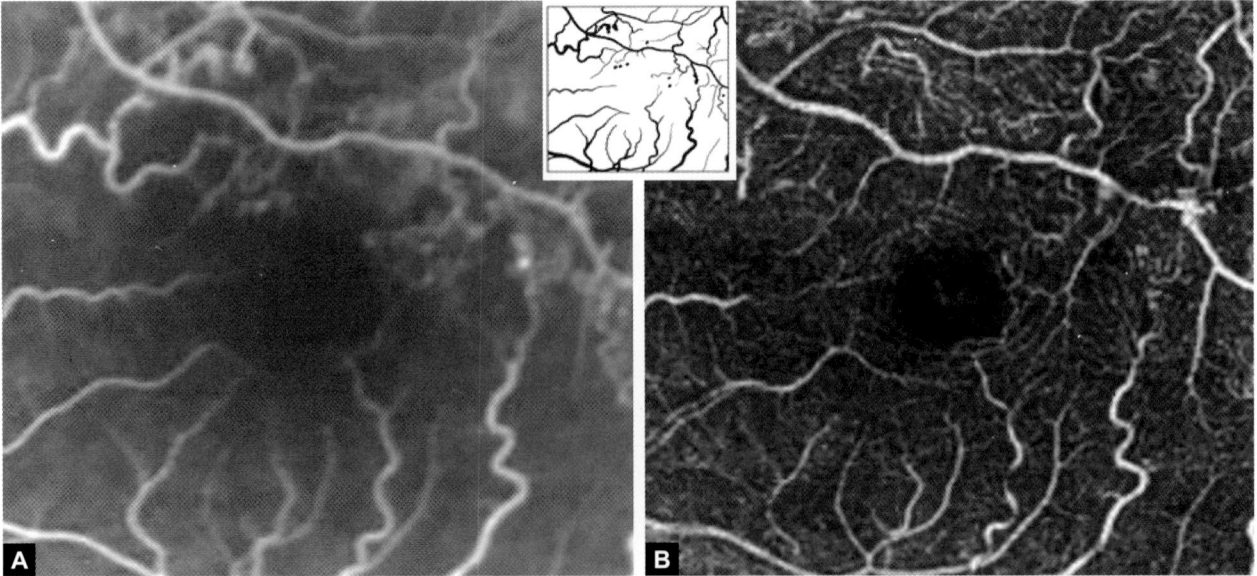

Figs 7.1.21A and B Branch vein occlusion. (A) Fluorangiography: Vessels with hyperfluorescent staining and areas of dye leakage are visible. There are also areas of ischemia; (B) Optical coherence tomography-angiography (Angio-OCT): Neither vessel staining, nor leakage; vessels show vascular flow. In the ischemic areas there is rarefaction of the vascular texture. In the areas of edema, the vessels are less regular and visible

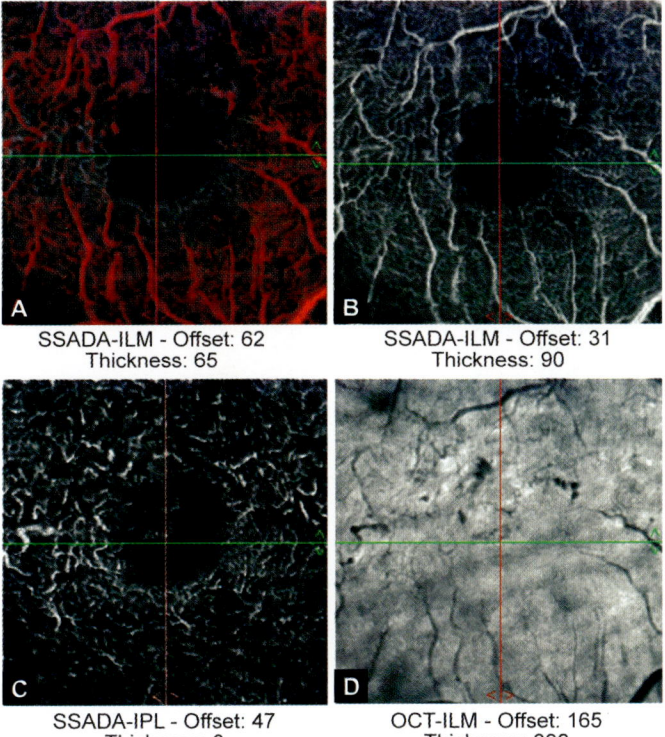

Figs 7.1.22A to D Branch vein occlusion. XR-Optovue screen shot of branch vein occlusion of the superotemporal retina of the fovea: (A) Inner/internal limiting membrane (ILM); the vascularization is highlighted in red; (B) Split-spectrum amplitude-decorrelation angiography (SSADA) image of the superficial vascularization; (C) SSADA aspect of deep vascularization at the inner plexiform layer; (D) En-face OCT image of ILM

Longstanding Retinal Ischemias

In the case of consolidated ischemias, when shunts are formed, it is possible to appreciate the small vascular branches that join the superficial plexus with the deep plexus. This aspect is appreciated dynamically by reducing the analysis of the superficial plexus with internal limiting membrane (ILM) to 30 μm and by proceeding in small steps towards the deep plexus. Small vascular trunks are seen that reach down to the deep plexus.

The changes in vessel size, flow and morphology of the plexus can be appreciated by analyzing the deep plexus. Within the deep plexus flow points can be observed and, by shifting the scan towards the surface, they are seen to join the shunt trunks. The shunt vessels evident flow could be due to the greater sensitivity of the split-spectrum amplitude-decorrelation angiography (SSADA) in highlighting the flow differences parallel to the incident light with respect to the transverse vascular structures (Figs 7.1.23 and 7.1.24).

DIABETIC RETINOPATHY

Background Retinopathy

In patients with background retinopathy, capillary nonperfusion areas, similar to the nonperfused areas highlighted by fluorangiography, are evident. Angio-OCT, however, shows a larger number of capillary loops and arteriovenous anastomoses. At the level of the deep capillary vascular plexus, the capillaries are more rarefied. Changes in size, vascular signal and morphology of the plexus are evident. Often the scarce capillaries are fan shaped. Also the anastomoses between superficial and deep vascular networks are very evident; these are not seen on the fluorangiography. Angio-OCT offers a much better view of anastomoses, especially the deep anastomoses and the vascular loops. The deep new vessels are more clearly seen than with angiography. Rare retinal hemorrhages are visible as masked areas, but they are less evident than they appear in fluorangiography.

Optical coherence tomography-angiography does not show up all the microaneurysms; those that are clearly seen are generally the larger microaneurysms where there is probably residual blood flow (Figs 7.1.25 to 7.1.27).

Advanced Retinopathy and Retinal Ischemia

The areas of retinal ischemia, examined with Angio-OCT are much sharper than as with fluorangiography because there is no masking effect by dye leakage. Details are appreciated that cannot be seen with fluorangiography because hidden by the dye in the intermediate and later stages of the examination.

Ischemic areas show that the capillaries are sparse and more evident against a gray background. Often the capillaries inside the nonperfusion areas are truncated, with abrupt interruptions, or with shunts, or anastomoses with the capillary layers of the deep vascular network. In Angio-OCT, the ischemic areas can be easily identified on the basis of texture and flow alterations. But there are some details in Angio-OCT that make it possible to distinguish the recent ischemic areas from the consolidated and chronic ischemic areas.

Initial neovascularizations are seen as thickened and irregular vessels that may emerge from the surface of the retina or the optic disk (Figs 7.1.28 to 7.1.41).

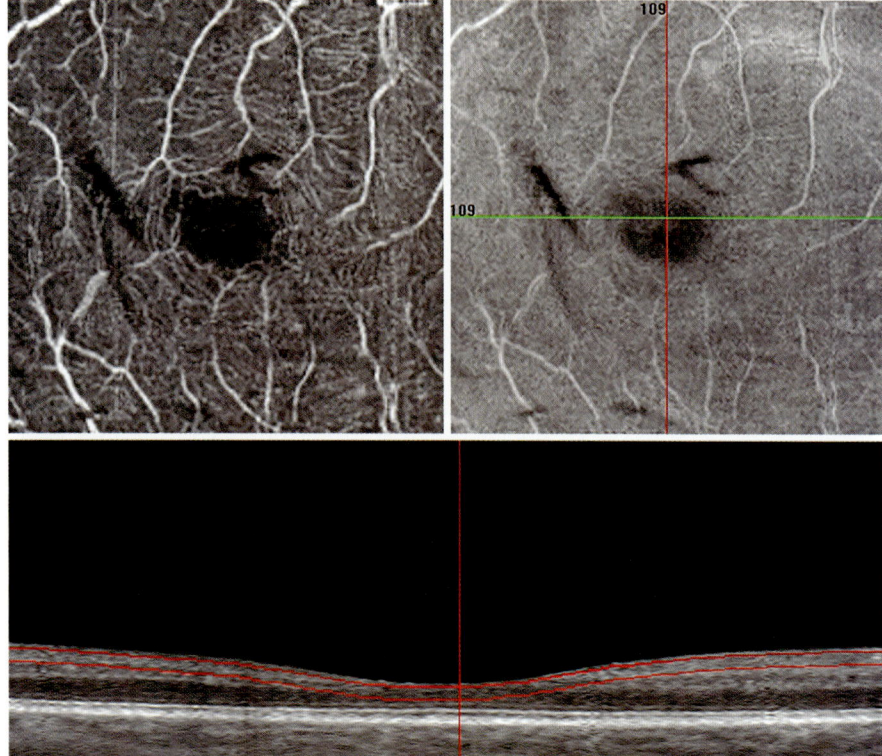

Fig. 7.1.23 Branch vein occlusion of superotemporal vein of the fovea, superficial vascular plexus. The occluded area presents rarefaction of the vascular texture. Some retinal ischemic areas are in contact with the avascular area. The ischemic area and the avascular area merged. Note screen effect formed by vitreous hemorrhages

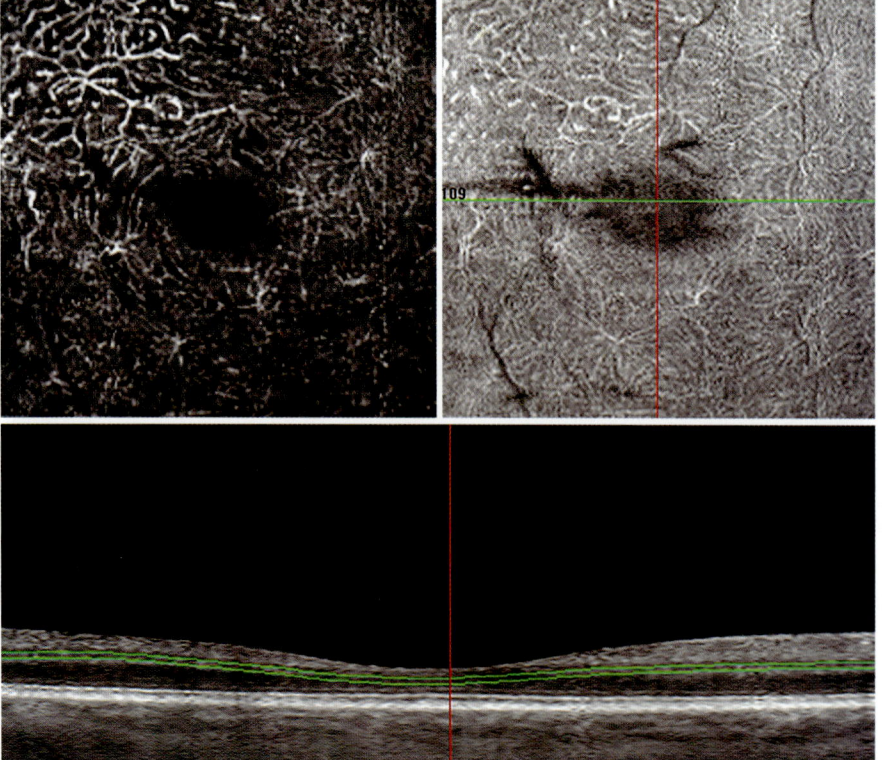

Fig. 7.1.24 Branch vein occlusion. Same case, the deep vascular plexus shows marked irregularity and fragmentation in the superotemporal area, while the rest of the retina appears to be normal

Clinical Applications: Aspects of OCT SSADA Angiography in Eye Disorders ■ 37

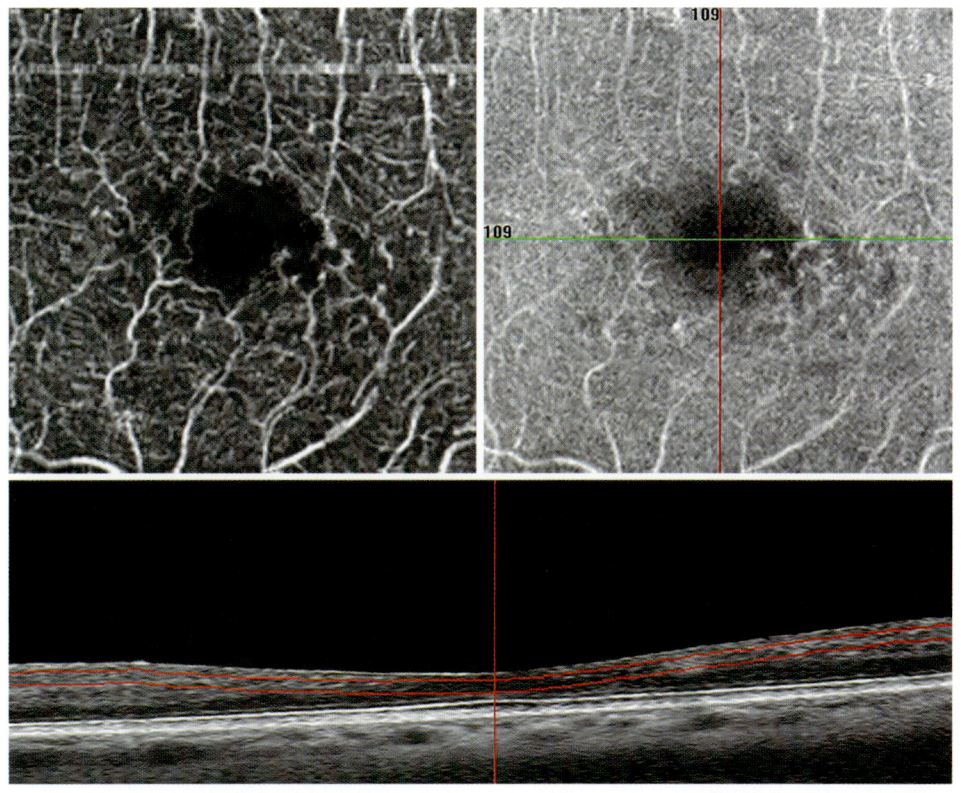

Fig. 7.1.25 Nonproliferative diabetic retinopathy: Optical coherence tomography-angiography (Angio-OCT) of the superficial vascular plexus. We see vascular network alterations, with capillaries more sparse. Some truncated capillaries, irregular perifoveal vascular network broken here and there, presence of microaneurysms

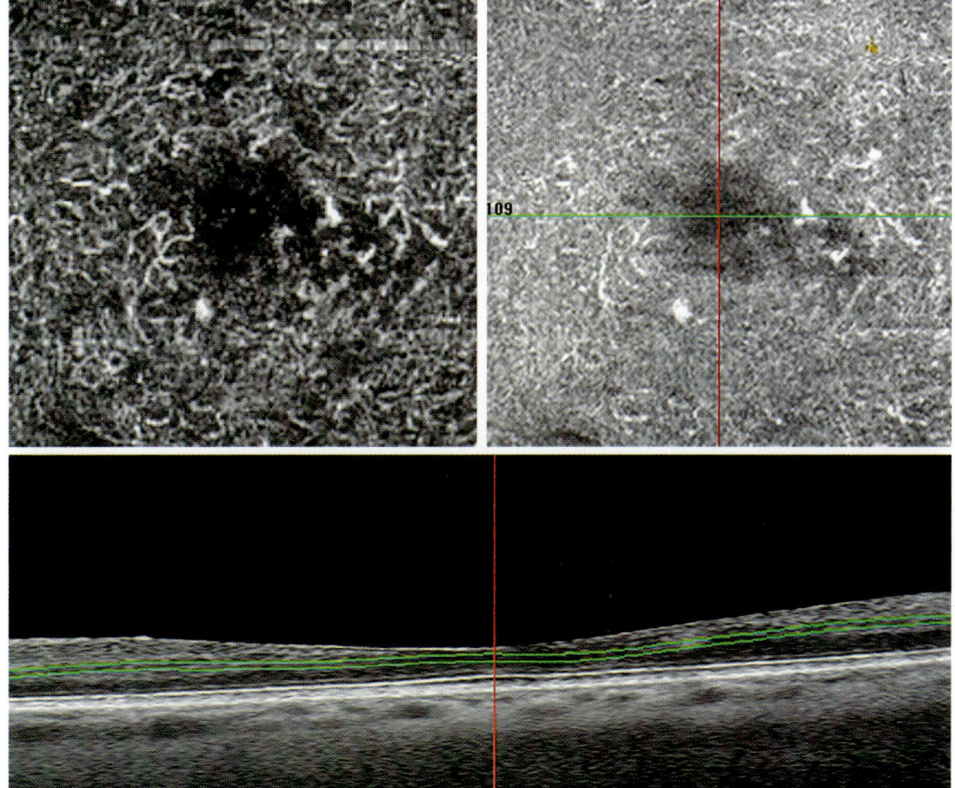

Fig. 7.1.26 Nonproliferative diabetic retinopathy: Optical coherence tomography-angiography (Angio-OCT) of the deep vascular network. Deep vascularization anomalous, with sparse capillaries and vascular fan pattern less rich in vascular branches. We see many microaneurysms and intraretinal microvascular alterations

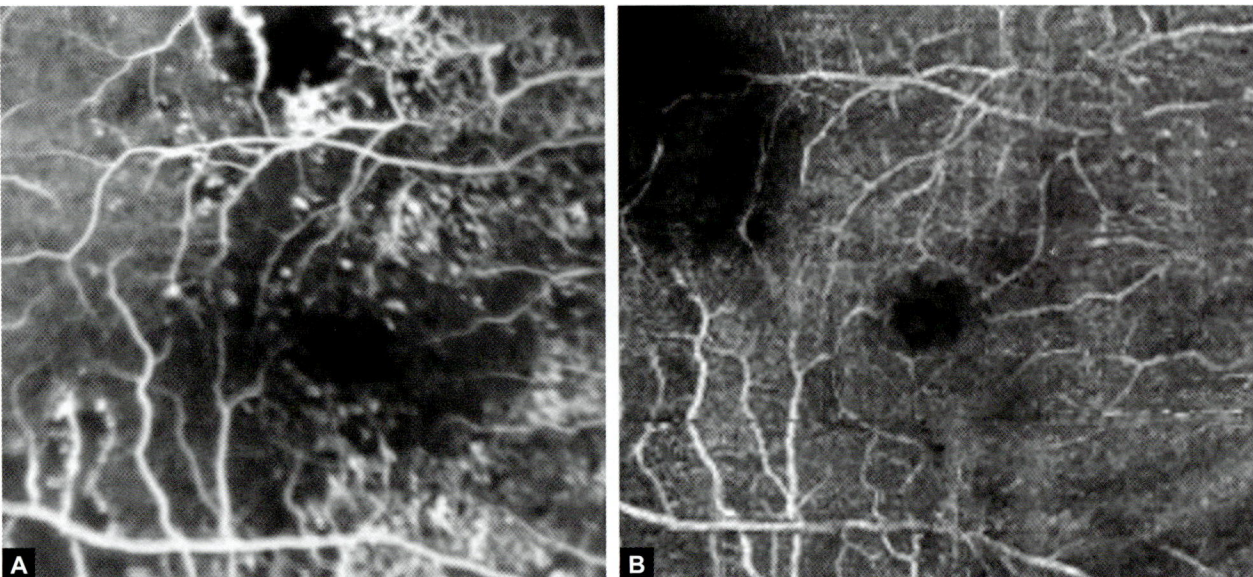

Figs 7.1.27A and B Diabetic retinopathy. (A) Fluorangiography highlights grayish ischemic areas, numerous microaneurysms, dilated vessels with stained walls; (B) Optical coherence tomography-angiography (Angio-OCT) of the same case; the microaneurysms are less visible and numerous. Vascular walls staining are not visible but around the vessel blood column a slight dark strip can be noticed that corresponds to the wall. Background texture is finely grained. In some points, the texture is coarser

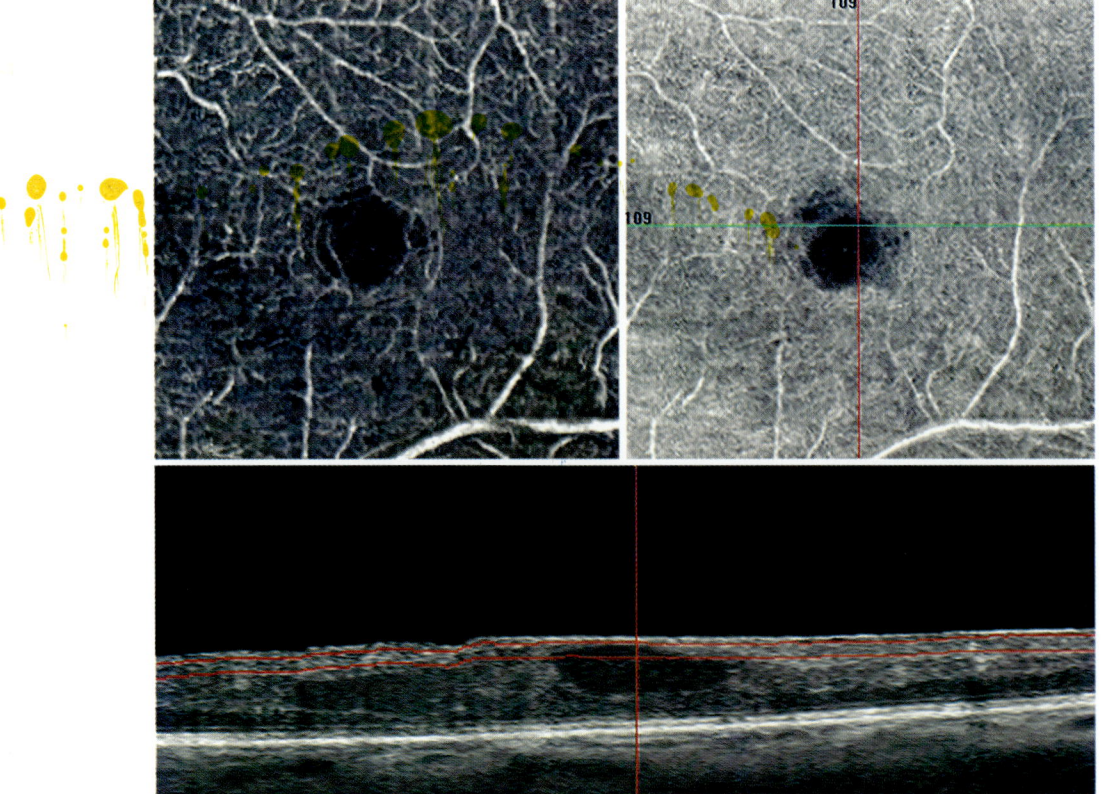

Fig. 7.1.28 Nonproliferative diabetic retinopathy, superficial vascular plexus. Optical coherence tomography-angiography (Angio-OCT) shows a vascular network less dense than in normal eyes and an enlarged avascular zone with some irregularities in the vascular arcade

Clinical Applications: Aspects of OCT SSADA Angiography in Eye Disorders 39

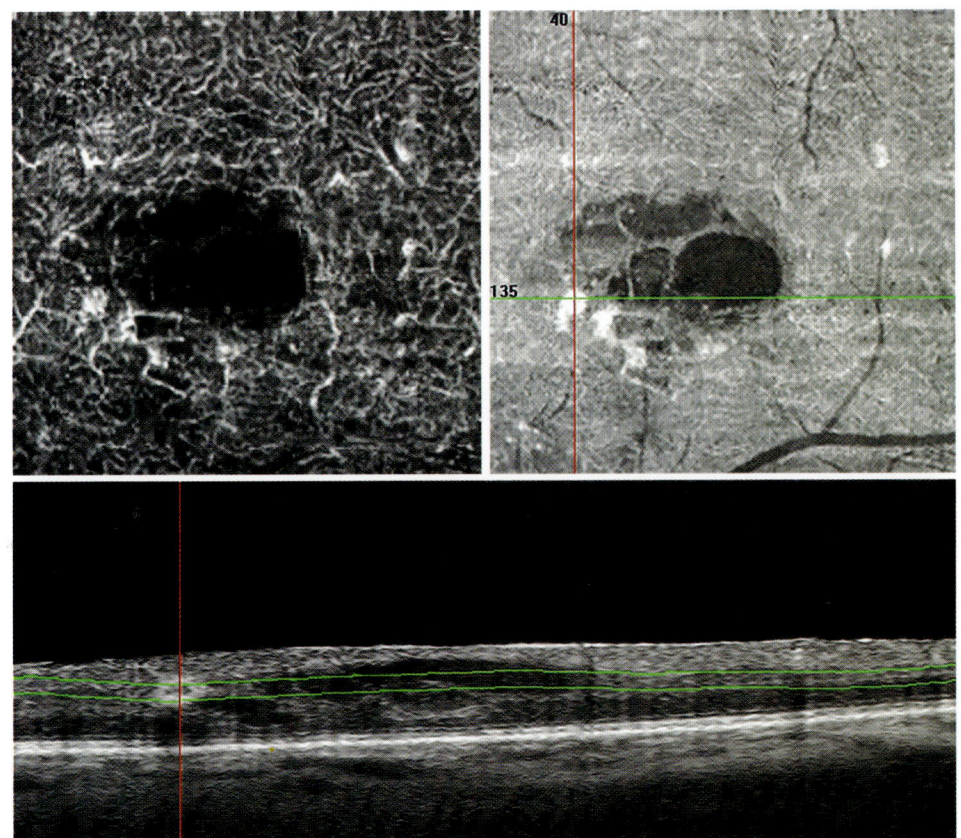

Fig. 7.1.29 Nonproliferative diabetic retinopathy, deep vascular plexus. This deep network, with fan-shaped vessels is less regular than in normal eyes; we can see dilatations, microaneurysms and intraretinal vascular anomalies

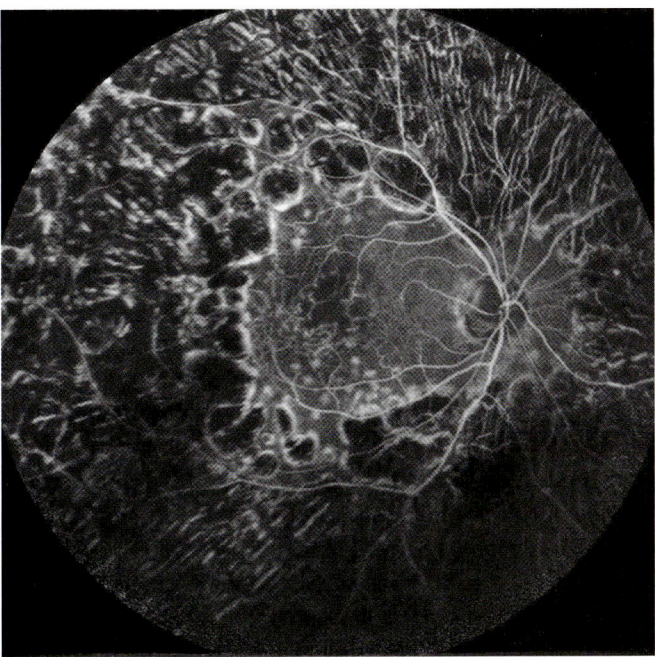

Fig. 7.1.30 Nonproliferative diabetic retinopathy with ischemic maculopathy; fluorangiography: the central areas of ischemia with central anomalies are evident. Note some laser scars around the macula

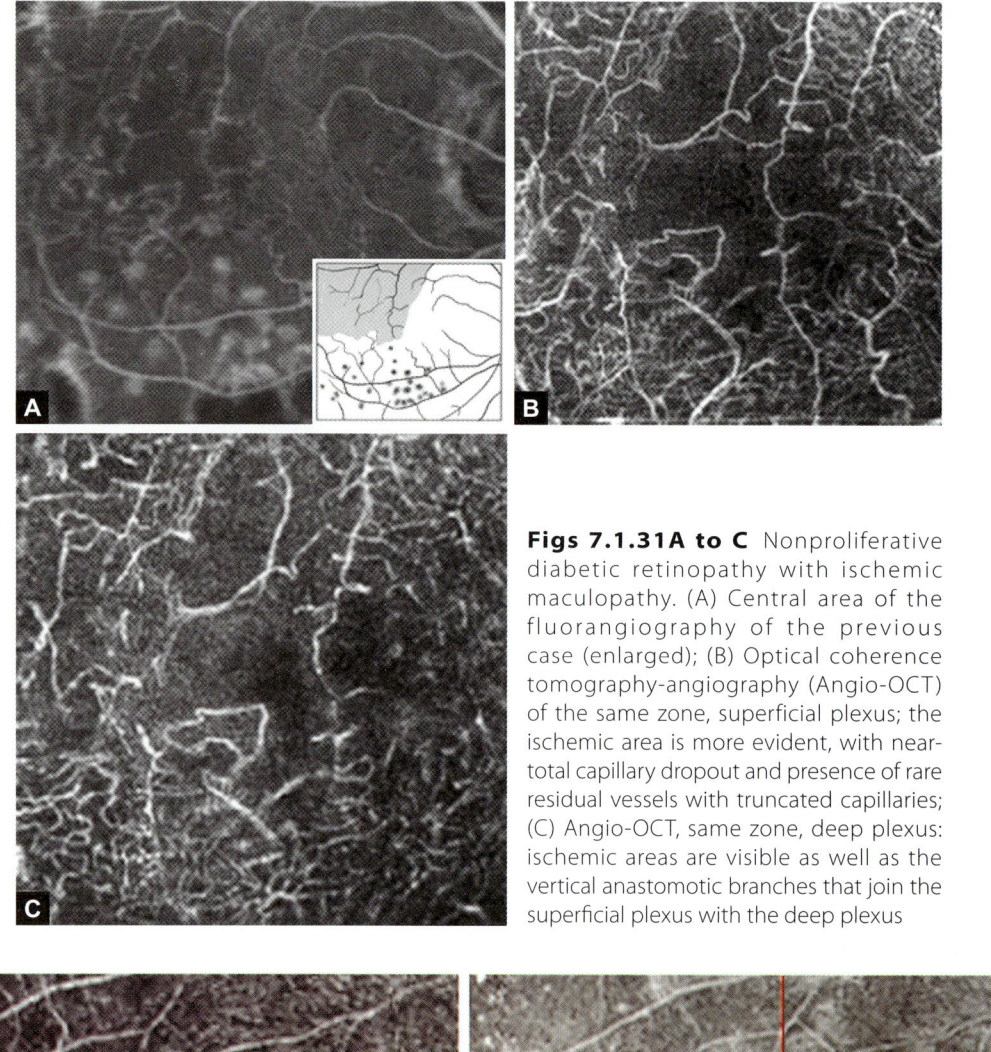

Figs 7.1.31A to C Nonproliferative diabetic retinopathy with ischemic maculopathy. (A) Central area of the fluorangiography of the previous case (enlarged); (B) Optical coherence tomography-angiography (Angio-OCT) of the same zone, superficial plexus; the ischemic area is more evident, with near-total capillary dropout and presence of rare residual vessels with truncated capillaries; (C) Angio-OCT, same zone, deep plexus: ischemic areas are visible as well as the vertical anastomotic branches that join the superficial plexus with the deep plexus

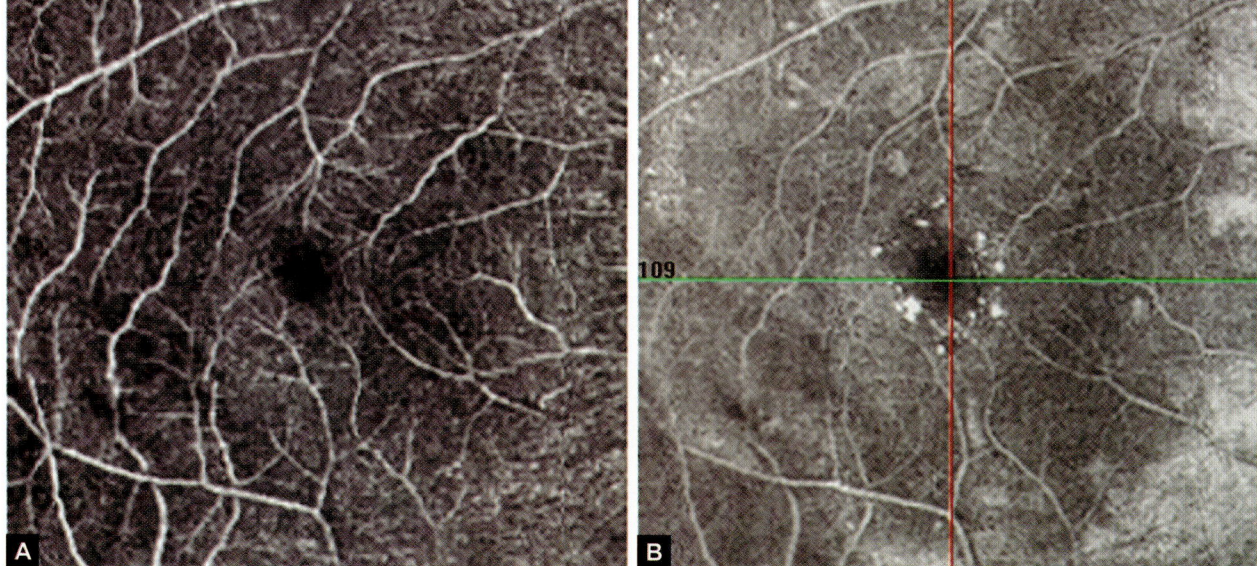

Figs 7.1.32A and B Nonproliferative diabetic retinopathy. (A) Optical coherence tomography-angiography (Angio-OCT) of the superficial plexus where the ischemic area is visible. Hard exudates that are present in the en-face image cannot be seen on Angio-OCT; (B) Classical en-face OCT of the same zone that highlights the hyper-reflectant perimacular exudates

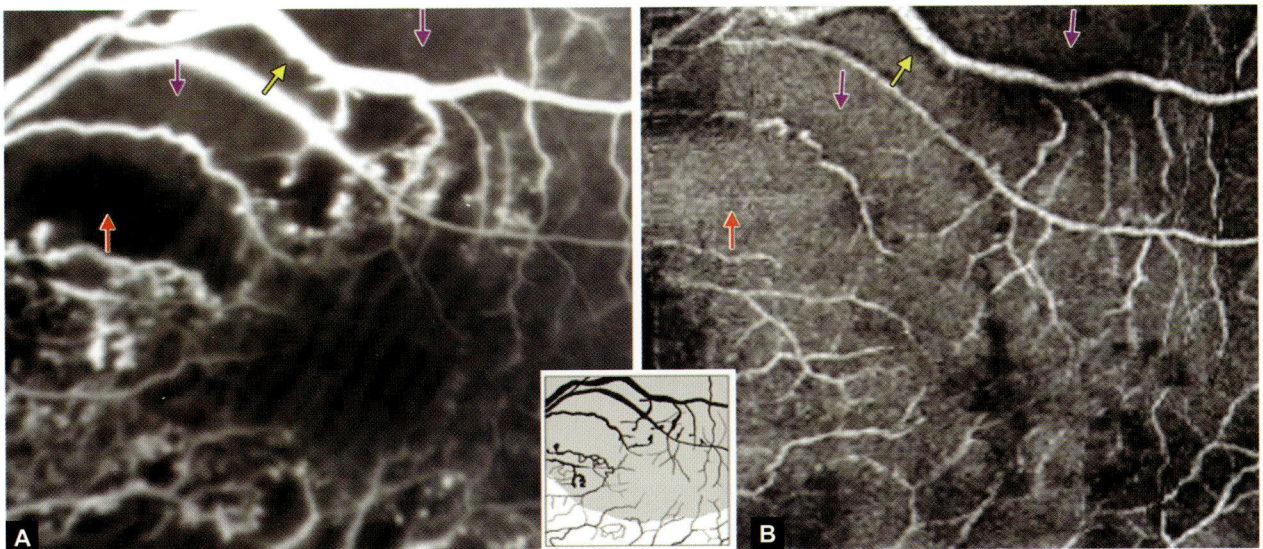

Figs 7.1.33A and B Diabetic retinopathy, ischemic area. (A) Fluorangiography: We can see dilated and markedly hyperfluorescent vessels. There is wall staining, with some leakage. The ischemic area is much darker than the lower areas where blood supply is normal; (B) Optical coherence tomography-angiography (Angio-OCT), same zone. Vessels are very thin compared to the fluorangiography, as Angio-OCT highlights only the blood flow. The vessel wall is seen as a darker area along the vessel course. The coarse granular texture at the posterior pole background can be appreciated while the figure's upper part shows an ischemic, darker and less granular texture

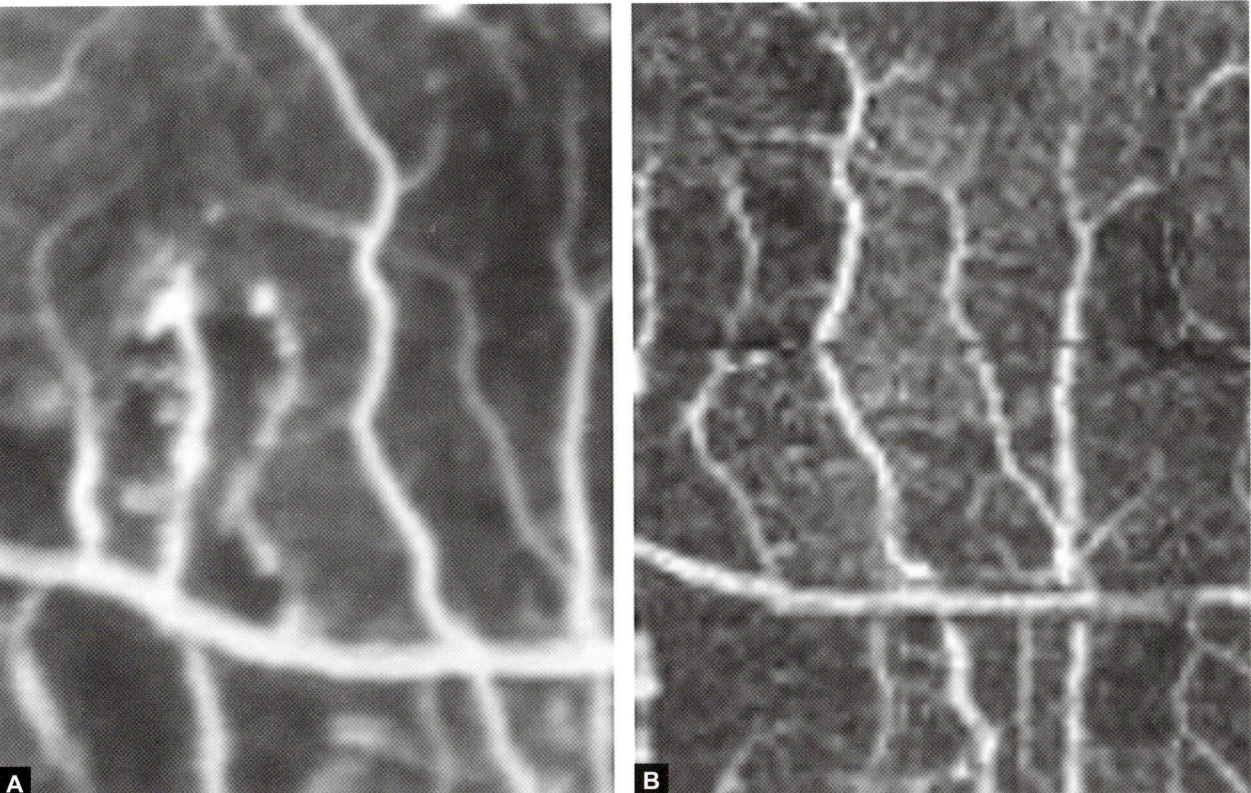

Figs 7.1.34A and B Diabetic retinopathy, ischemic form. (A) Fluorangiography with hyperfluorescent vessels and nonperfused areas; (B) Optical coherence tomography-angiography (Angio-OCT). The vessels are much thinner with dark borders, background texture is granular and coarse

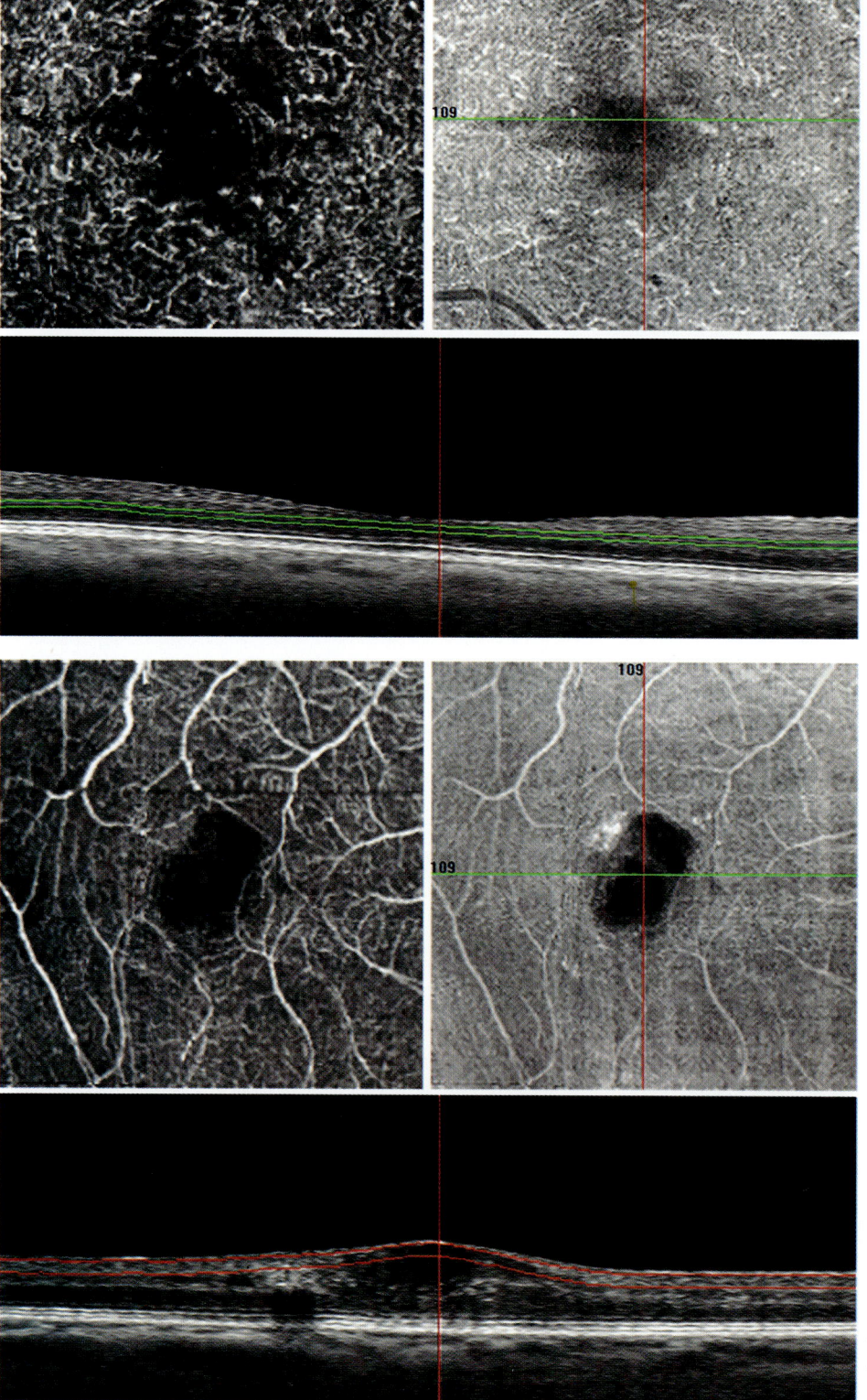

Fig. 7.1.35 Diabetic retinopathy deep vascular plexus. Disruption and major irregularity in the deep network with uneven capillaries, smaller and more sparse vascular fans, microaneurysms with intraretinal microvascularization

Fig. 7.1.36 Nonproliferative diabetic retinopathy with marked vascular anomalies at the superficial vascular plexus. The plexus is irregular with trapezoid avascular areas and interruptions in the vascular arcades; presence of small cystoid edema cells

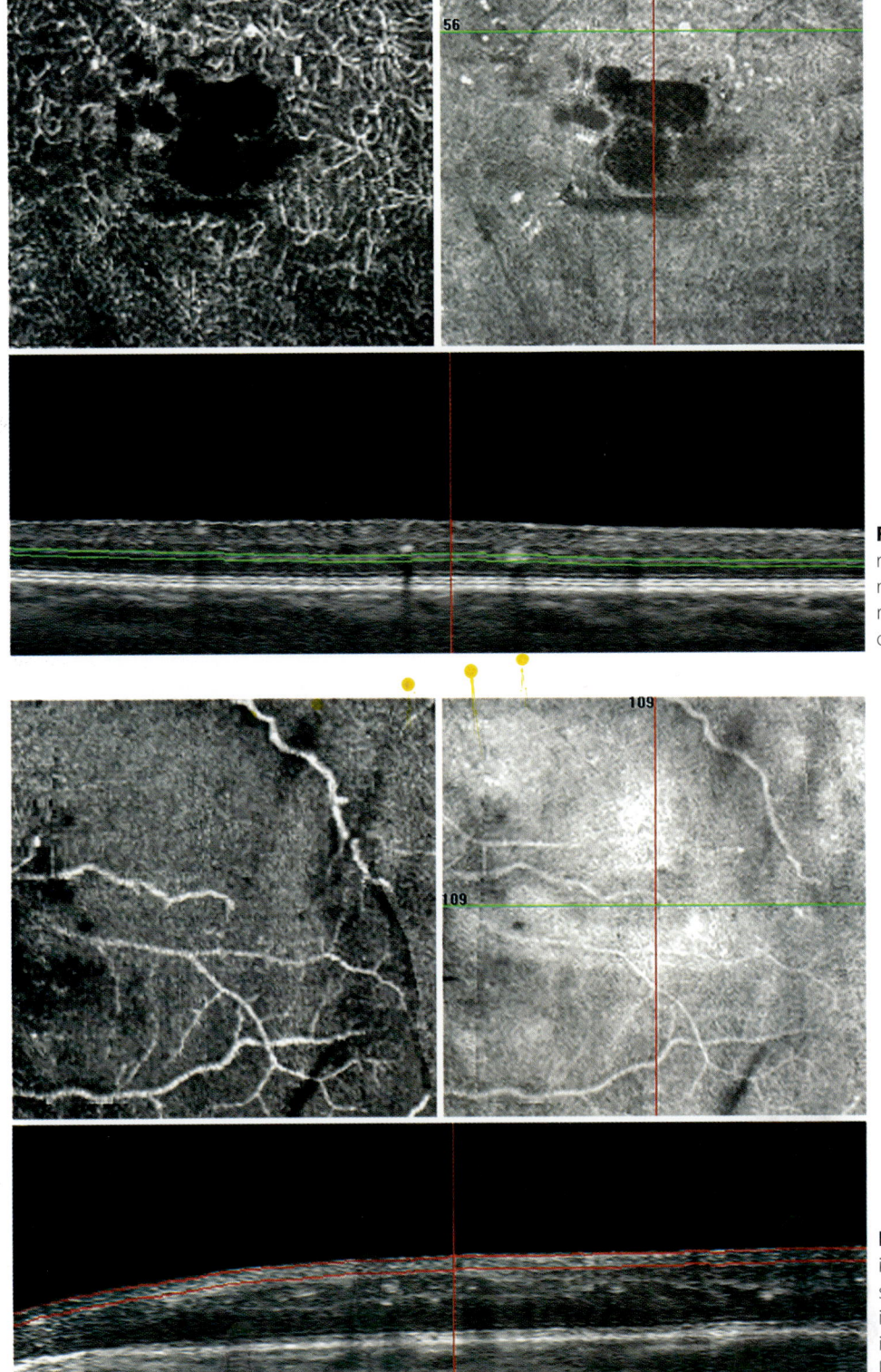

Fig. 7.1.37 Nonproliferative diabetic retinopathy: same eye, deep vascular network, evident microvascular alterations around the small cystoid edema cells

Fig. 7.1.38 Diabetic retinopathy, ischemic area. Figures 7.1.38 to 7.1.41 show various intraretinal layers of an ischemic diabetic retinopathy, in an ischemic area. A vertical anastomotic branch between upper and deep plexuses can be appreciated

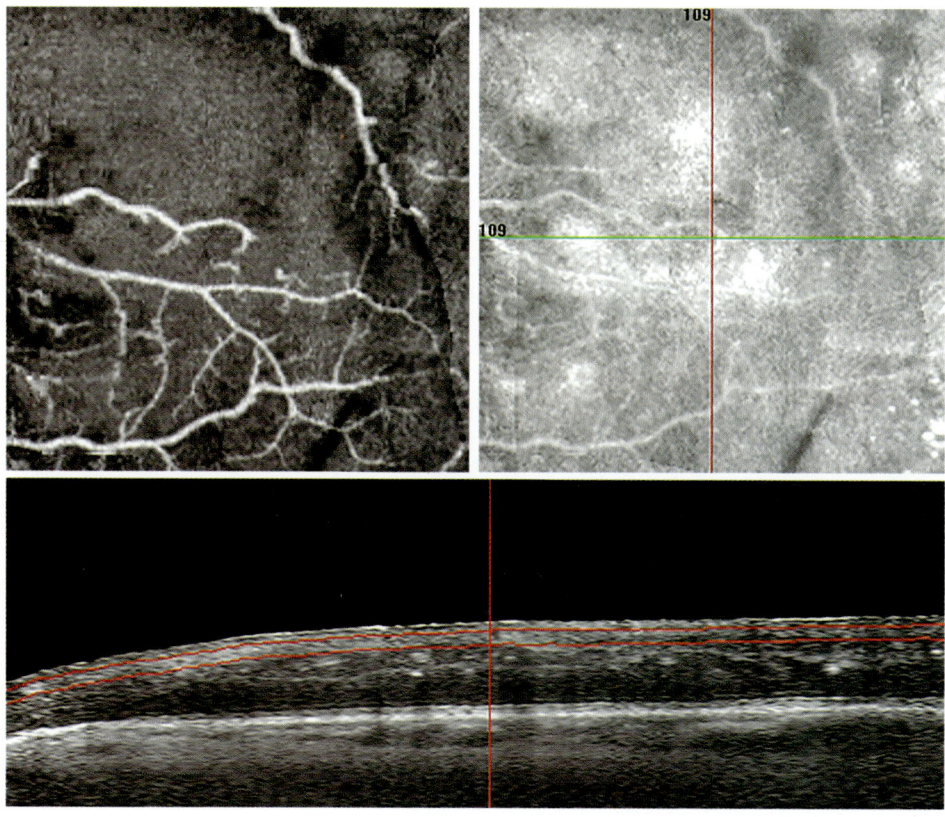

Fig. 7.1.39 Superficial scan: note the ischemic area with a very regular and very fine texture

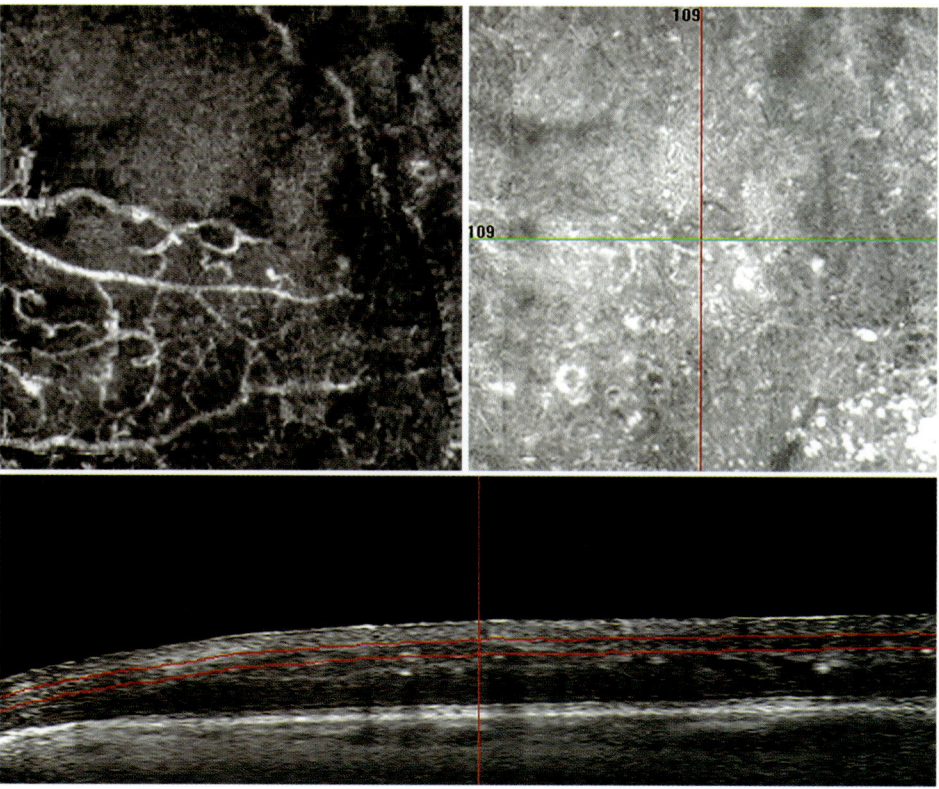

Fig. 7.1.40 Deeper scan that highlights the disruption of the deep vascular plexus

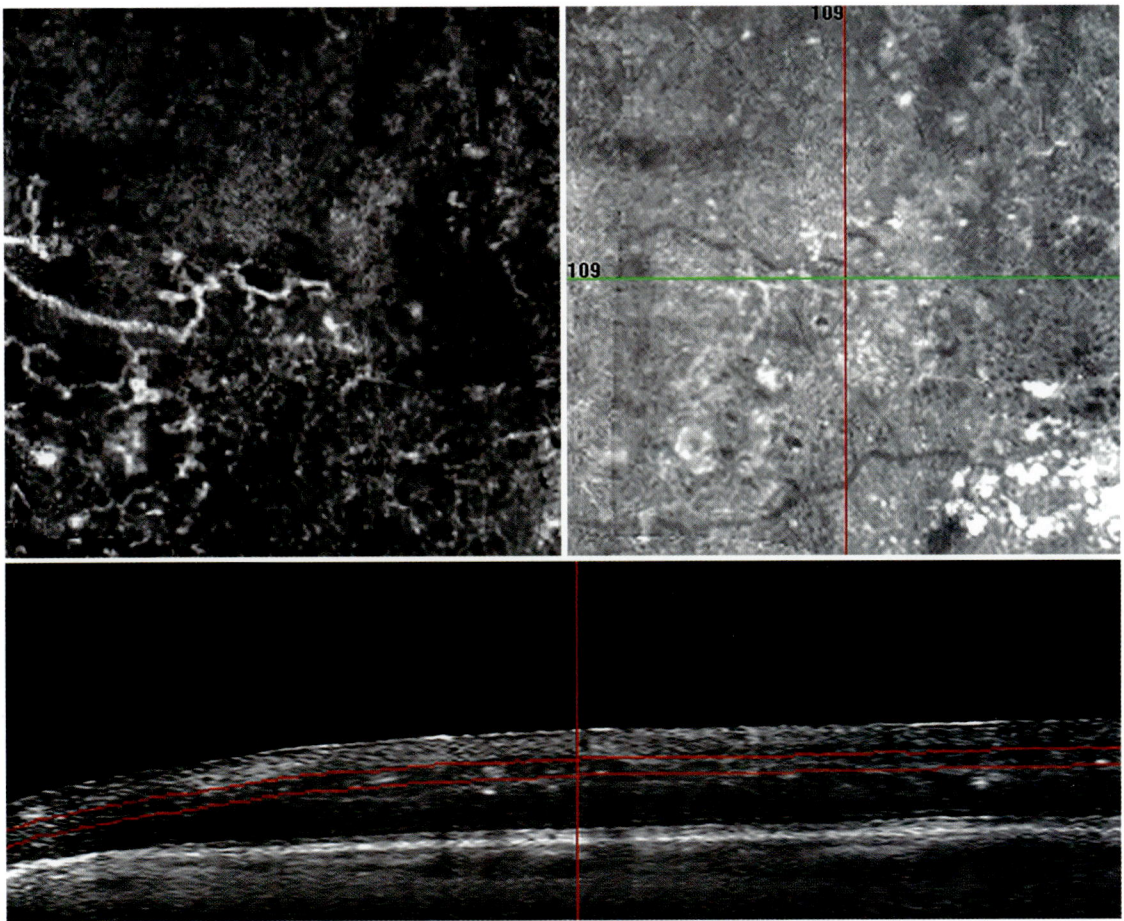

Fig. 7.1.41 Deeper scan

PROLIFERATIVE DIABETIC RETINOPATHY

The natural evolution of ischemic area in diabetic retinopathy, or in ischemic venous occlusions, is characterized by the progressive formation of new vessels, preceded by the establishment of capillary shunts that are clearly visible in fluorangiography. Indeed, fluorangiography shows up the newly formed vascular loops of the artery-artery and arteriovenous types. With fluorangiography however it is not possible to appreciate the level of these alterations, but only the two-dimensional course, the veins dilatation and the diffusion of the newly formed capillaries.

In diabetic retinopathy, chronic ischemia leads to proliferative diabetic retinopathy with preretinal and prepapillary neovascular membranes. Initial new vessels are seen as thickened and irregular capillaries that may emerge from the surface of the retina or the optic disk. In fluorangiography there is a very intense dye leakage that does not allow us to see the neovascularization.

Optical coherence tomography-angiography of preretinal and prepapillary neovascular membranes allows the operator to make a very precise evaluation of the extent and morphology of the network without the problems linked to dye leakage. The flow and morphology of the neovascular network is perfectly visible. Angio-OCT can be performed during pregnancy and allows us to follow evolution after laser pan retinal photocoagulation (Figs 7.1.42 and 7.1.43).

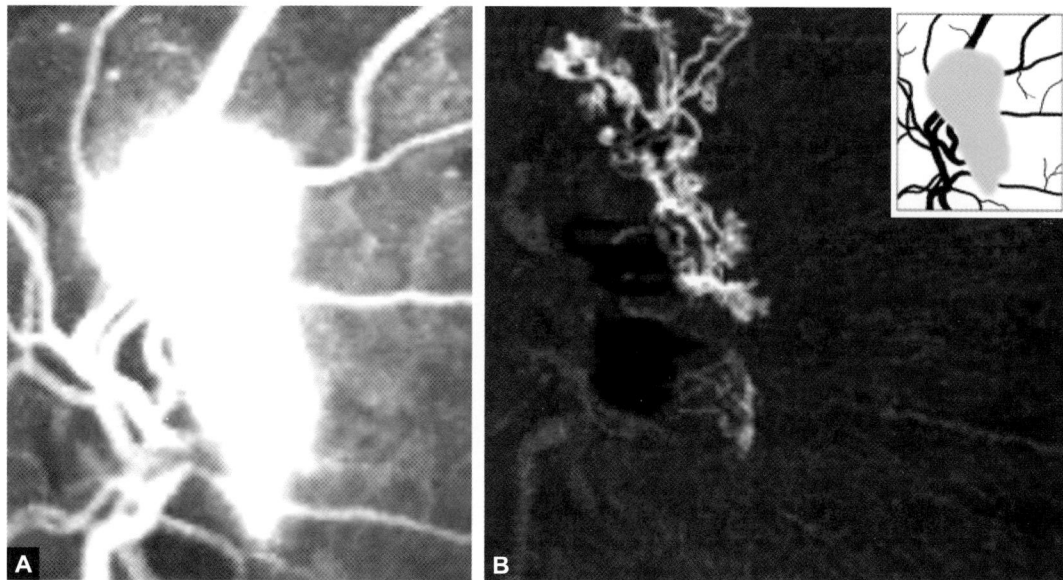

Figs 7.1.42A and B Proliferative diabetic retinopathy with prepapillary neovascular membrane at the optic disk. (A) Fluorangiography: The main retinal dilated hyperfluorescent vessels around the papilla are clearly visible, while the prepapillary vascular network is hidden by a dense hyperfluorescent cloud due to dye leakage; (B) The optical coherence tomography-angiography (Angio-OCT) shows the same area. Since there is no leakage of the contrast medium, the prepapillary new formed vessels are quite visible, threadlike and tangled in some points. Blood flow is more evident in some capillaries, less evident in other. Angio-OCT shows the blood flow inside the new formed vessels

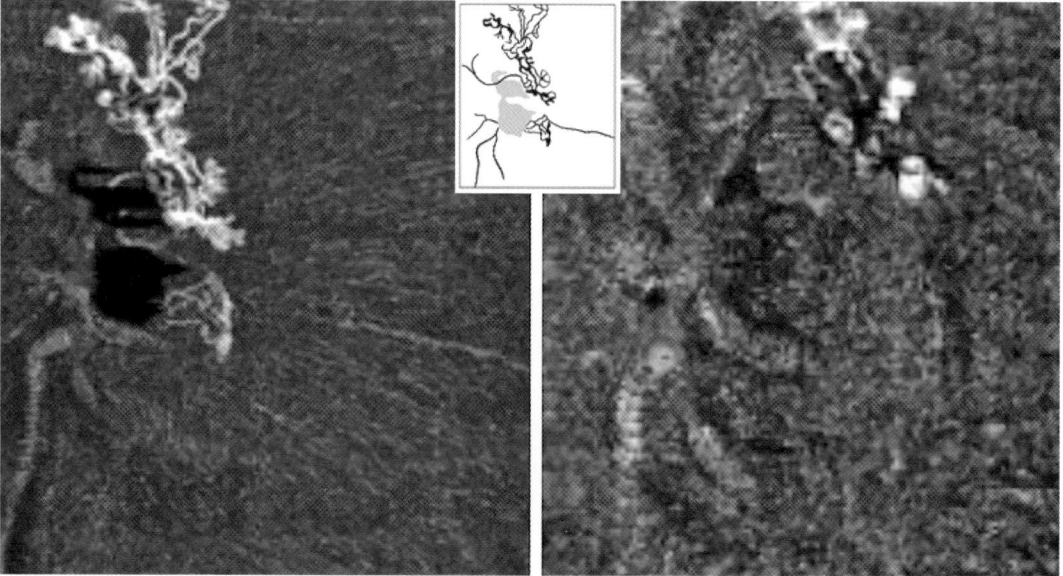

Fig. 7.1.43 Proliferative diabetic retinopathy with neovascular prepapillary membrane at the optic disk; same patient. An intense panretinal treatment caused an important regression of the new vessels. Some fragmented vessels broken down into three tangled residual formations still persist left, before treatment, right, after treatment

PART II — OUTER RETINA

CNV: NEOVASCULAR MEMBRANES IN ARMD

The Angio-OCT of neovascular membranes allows the operator to make a very precise evaluation of the extent and morphology of the network without the problems linked to the dynamics of the dye. Inside the fibrosis tissue the flow and morphology of the neovascular network is always visible.

Study of the subretinal neovascularization in fluorangiography provides very important data on the progress of the pathology. The new vessels leak the dye, masking rapidly the lesion morphology. The neovascular network is seen only in the early angiographic stages and only for a very few seconds. Following treatment of these lesions with anti-vascular endothelial growth factor (VEGF), it is difficult to precisely observe treatment efficacy because the new vessels stain intensely, even though they could significantly regress. Finally within a fibrovascular membrane, staining totally masks the neovascular network even when it is greatly reduced by the treatment.

The Angio-OCT of neovascular membranes allows us to be very precise in evaluating the extent and morphology of the network without the problems linked to the dynamics of the dye. The morphology of the neovascular network inside the fibrovascular formation is always visible. Angio-OCT in the follow up of intravitreal treatments allows us to observe the regression of the new vessels and even their disappearance.

New Vessels Morphology

Some neovascular membranes have a tree-like aspect, with fine ramifications that at times seem to infiltrate the subretinal and retinal tissues. The shape of the new vessels may be similar to a fan, octopus and spiderweb. These evident vascular formations (flows) have an irregular and faint aspect that is totally different from the normal retinal and choroidal vascular networks. The new vessels are thin and irregular. They are at times, observed inside nonvascular connective tissue. A feeder trunk can almost always be noticed, or a bunch of feeder vessels, as well as the peripheral anastomoses.

Optical coherence tomography-angiography may show the neovascular membranes at times, as cartwheels or bicycle wheels with anastomoses of the peripheral branches that have the peculiarity of being located only within the deep retinal layers above the pigment epithelium (type II), or below retina pigment epithelium (RPE) and above Bruch's membrane (BM) (type I) above the choroid. The new vessels are thin and irregular, inside the fibrovascular tissue. We can see almost always a feeder vessels as well as peripheral anastomoses.

The fluorangiography aspect is similar in the early part of the examination for only a few seconds, while it is very different in the intermediate and late stages because of the almost immediate masking effect produced by dye leakage.

Fibro Vascular Formation

In the more advanced forms, the Angio-OCT shows small irregular vascular networks inside a fibro vascular formation. The networks are thin and irregular and are observed inside nonvascular tissue. In this case, the OCT horizontal scan will need thickness necessary to detect the flows inside the fibrosis.

Post-treatment Flows

In our limited experience, the new vessels apparent regression almost always precedes the reabsorption of the subretinal or subepithelial fluid and of the edema. In the more advanced forms, after repeated treatment, the Angio-OCT highlights images of small masked vascular networks inside a capsular formation. The networks are thin and irregular and are observed inside the connective tissue. In order to obtain a good image, the segmentation will have to include the entire fibrotic area with a thickness sufficient to detect flows inside the thickness of the fibrosis.

After treatment with anti-VEGF, a partial regression of the new vessels, and even their disappearance, can be observed. On the day following the injection of anti-VEGF, almost all the neovascular branches disappear, showing only thinned and sparse residual branches. The vascular network, however, is again visible after 7 days. These are early observations that need to be confirmed on a larger number of patients. We have observed various types of responses to therapy, due both to the efficacy of the therapeutic substance or to the chronic nature of the disorder.

Longstanding Fibrosis

Even in the case of longstanding fibrosis, Angio-OCT may provide images of small vascular networks inside the fibrotic scar. We do not know whether these are quiescent new vessels or residual channels, with a minimum amount of blood flow. In old scars, Angio-OCT highlights small vascular networks masked by the fibrous tissue of the fibrotic scar. Segmentation will have to pass through the fibrotic area with a thickness sufficient to detect the flows inside the fibrosis. In some advanced cases, the rarefied capillaries appear to be stable channels (Figs 7.1.44 to 7.1.63).

NEOVASCULAR MEMBRANES IN MYOPIC EYES

Optical coherence tomography-angiography highlights neovascular membranes in myopic eyes as irregular close-knit flow formations observed at the level of the deeper retina layers, in contact with RPE. Subretinal myopic new vessels are thin and irregular, at times with a glomerular aspect, apparently contained inside a very thin capsular formation. These vascular flows are absolutely different from the retinal and choroidal vascular networks.

After injection of anti-VEGF the neovascular network disappears almost entirely. It may reappear only partially after 7 days. These observations need to be confirmed on a larger number of cases. The fluorangiography is immediately masked by dye leakage (Figs 7.1.64 to 7.1.68).

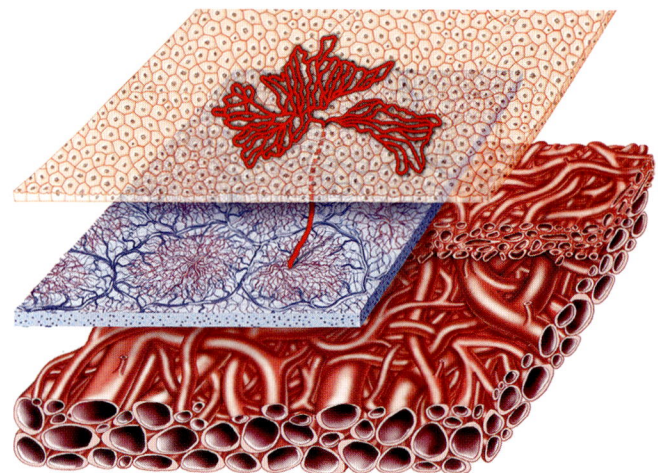

Fig. 7.1.45 Neovascular classical membrane, [choroidal neo-vascularization (CNV) type 2]; The classical neovascular membrane, or type 2 CNV membrane, develops above the pigment epithelium, underneath the retina; it may infiltrate the retina. It too arises from the choriocapillaris, with a feeder vessel. The type 2 membrane always determines serous elevations of the retina and cystoid edema

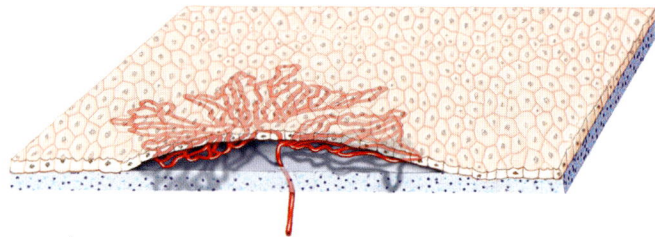

Fig. 7.1.46 Neovascular occult membrane, [choroidal neo-vascularization (CNV) type1], pigment epithelium vascularized detachment. The membrane originates from the choriocapillaris, with a feeder vessel that perforates Bruch's membrane (BM) and develops under the pigmented epithelium forming initially some vascular loops that later become vascular fans. The pigmented epithelium is elevated by the occult new vessels that initially form a simple vascular network

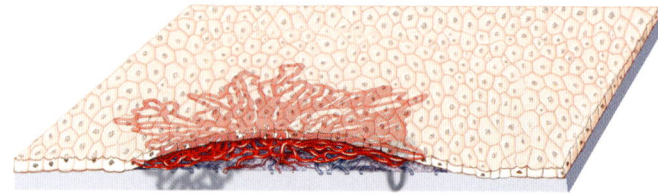

Fig. 7.1.47 Neovascular occult membrane, [choroidal neo-vascularization (CNV) type1], pigment epithelium vascularized detachment. The pigment epithelium is elevated by the occult new vessels. The membrane initially forms some vascular loops that later become dense vascular fans. These fans proliferate inside the pigmented epithelium detachment, until they fill it

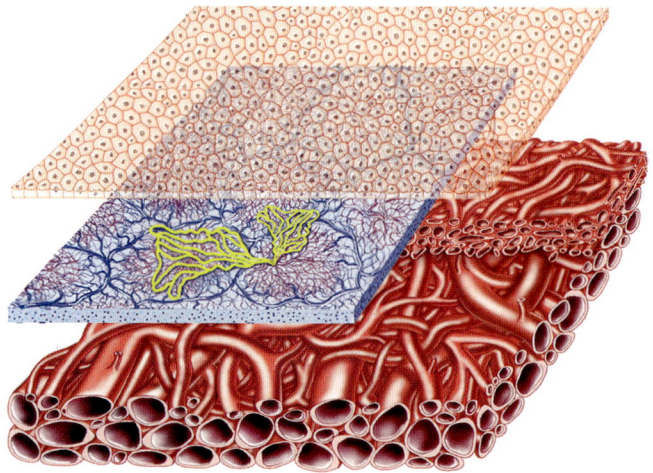

Fig. 7.1.44 Neovascular occult membrane, [choroidal neo-vascularization (CNV) type 1]. The membrane originates from the choriocapillaris, with a feeder vessel that perforates Bruch's membrane (BM) and develops under the pigment epithelium forming initially some vascular loops that later become vascular fans

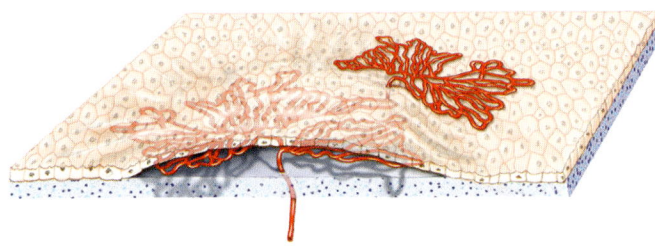

Fig. 7.1.48 Neovascular occult membrane [choroidal neovascularization (CNV) type1], pigment epithelium vascularized detachment. The pigment epithelium is detached from Bruch's membrane (BM), forming a vascularized detachment. The vascular fans proliferate inside the pigmented epithelium detachment, until they completely fill it. The pigment epithelium is then perforated in one or more points by the new vessels that will develop above the pigment epithelium, underneath the retina that is elevated and infiltrated by them. This active membrane always determines serous elevations of the retina and cystoid edema

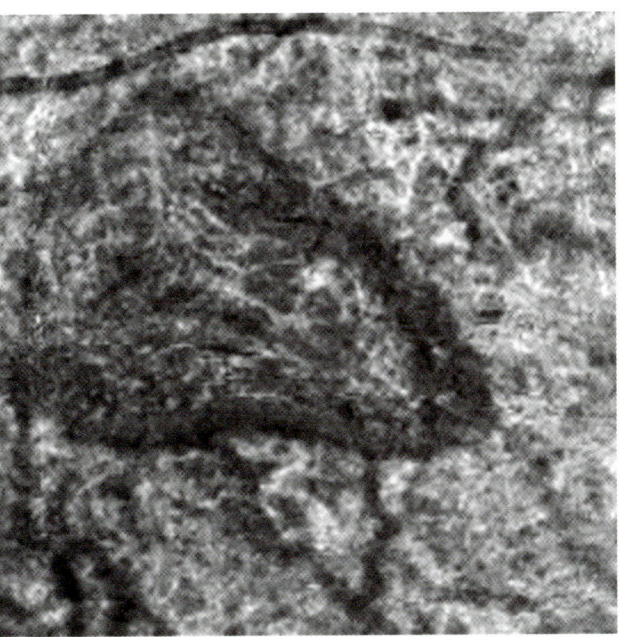

Fig. 7.1.50 Age-related macular degeneration (AMD). Same case 24 hours after a new injection of Aflibercept. The blood flow appears less dense, more fragmented, with loss of most capillaries

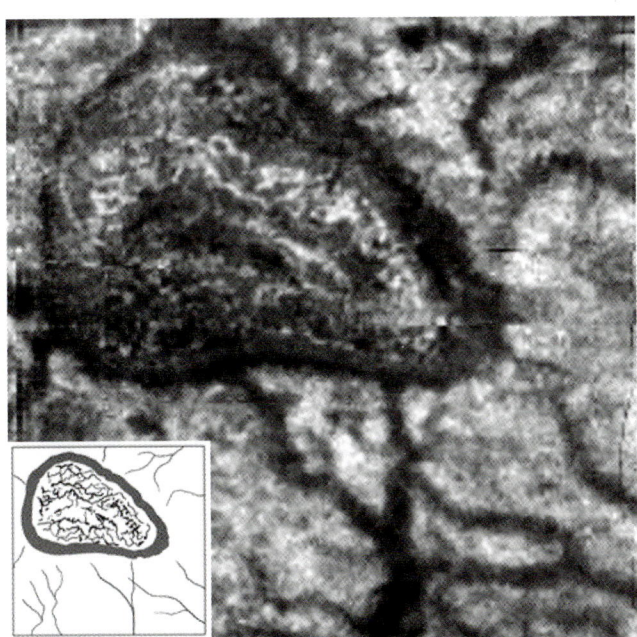

Fig. 7.1.49 Age-related macular degeneration (AMD), relapsing neovascular membrane after three injections of Aflibercept. The neovascular membrane is observed inside a fibrovascular formation where the vessels blood flow form a fragmented and dense irregular network, wholly contained inside the fiberoptic formation

IDIOPATHIC POLYPOIDAL CHOROIDAL VASCULOPATHY

Optical coherence tomography-angiography may show a vascular net between angiomas, at RPE level. They appear as flow formations at the level of the deeper retina layers, in contact with RPE (Figs 7.1.69A and B).

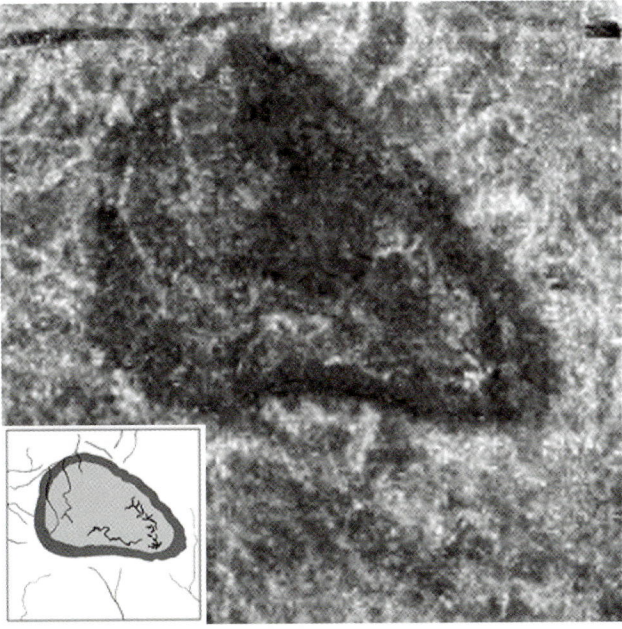

Fig. 7.1.51 Age-related macular degeneration (AMD). Same case, 7 days after Aflibercept injection. The neovascular network decreased further and is even less visible. Only a few fragmented branches are observed

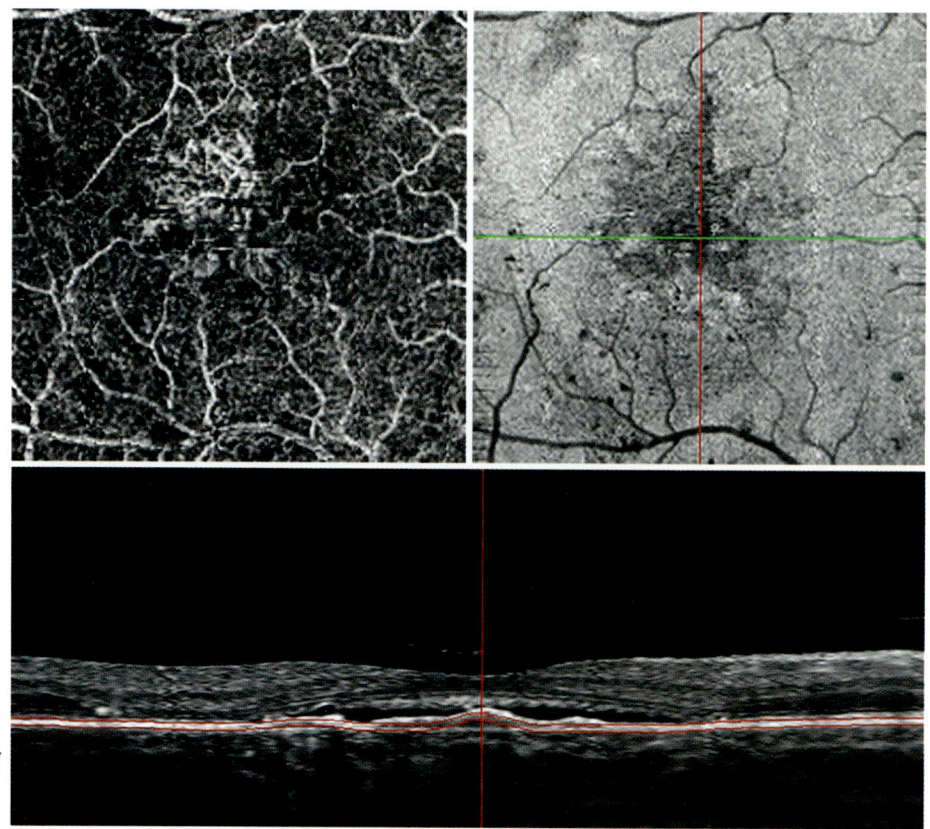

Fig. 7.1.52 Age-related macular degeneration (AMD). Occult neovascular membrane. A dense tree-like neovascular network is observed underneath the pigment epithelium flat detachment. The margins are not evident. However, the entire network is contained in the detached area

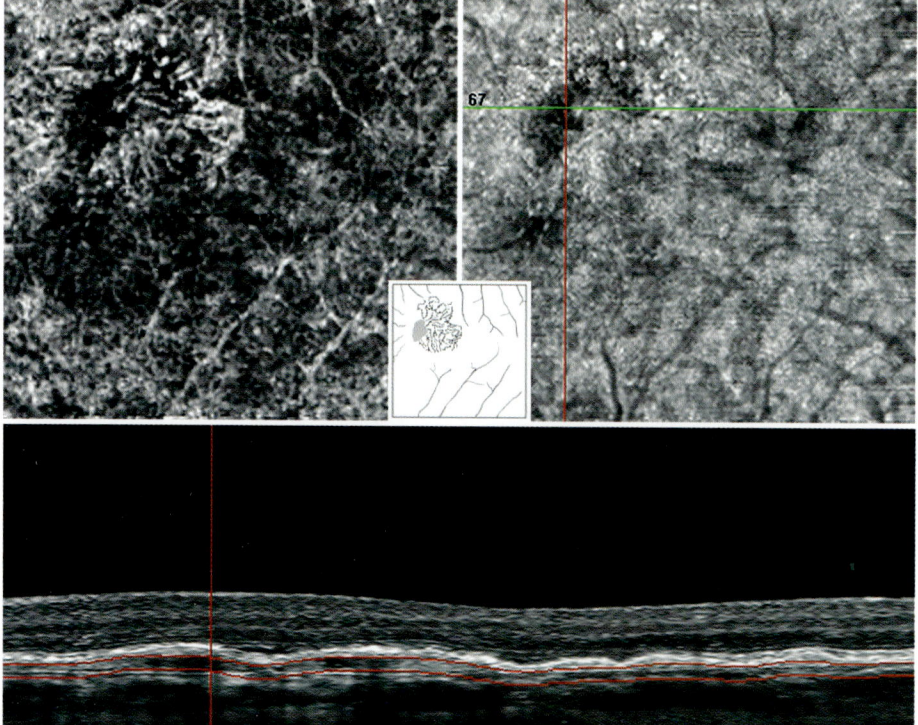

Fig. 7.1.53 Age-related macular degeneration (AMD). Occult neovascular membrane after treatment. The neovascular membrane is less dense and more fragmented. There are still evident signs of activity, namely capillaries containing blood flow

Clinical Applications: Aspects of OCT SSADA Angiography in Eye Disorders   51

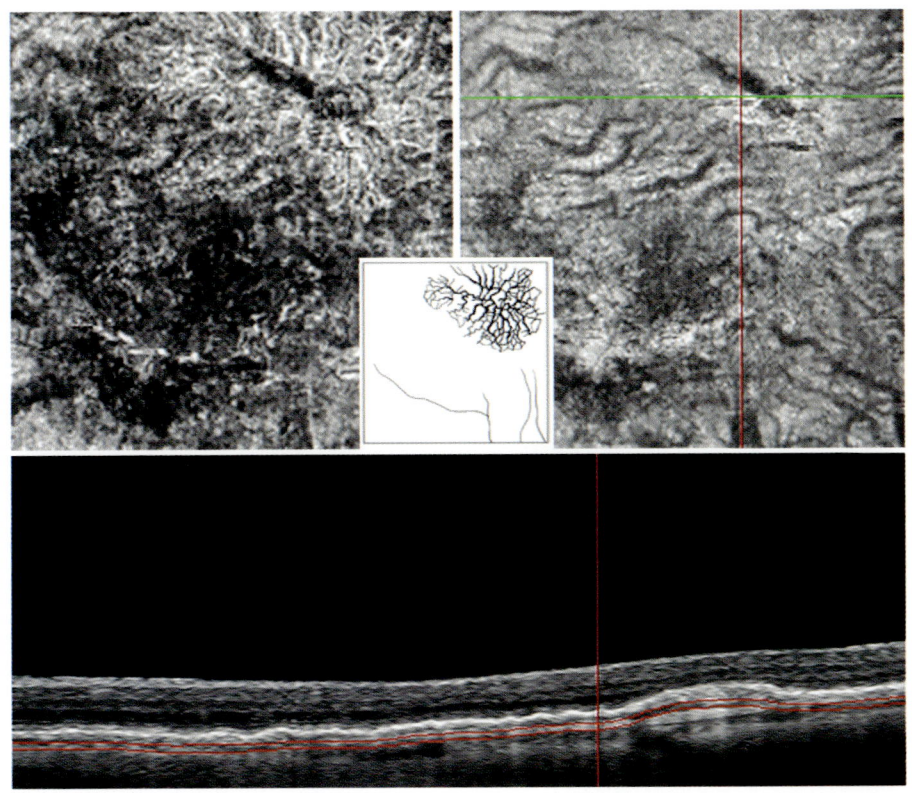

Fig. 7.1.54 Age-related macular degeneration (AMD): The neovascular membrane that has relapsed several times and is still active. We see a tree-like pattern with many capillaries. Blood flow is seen but not very evident

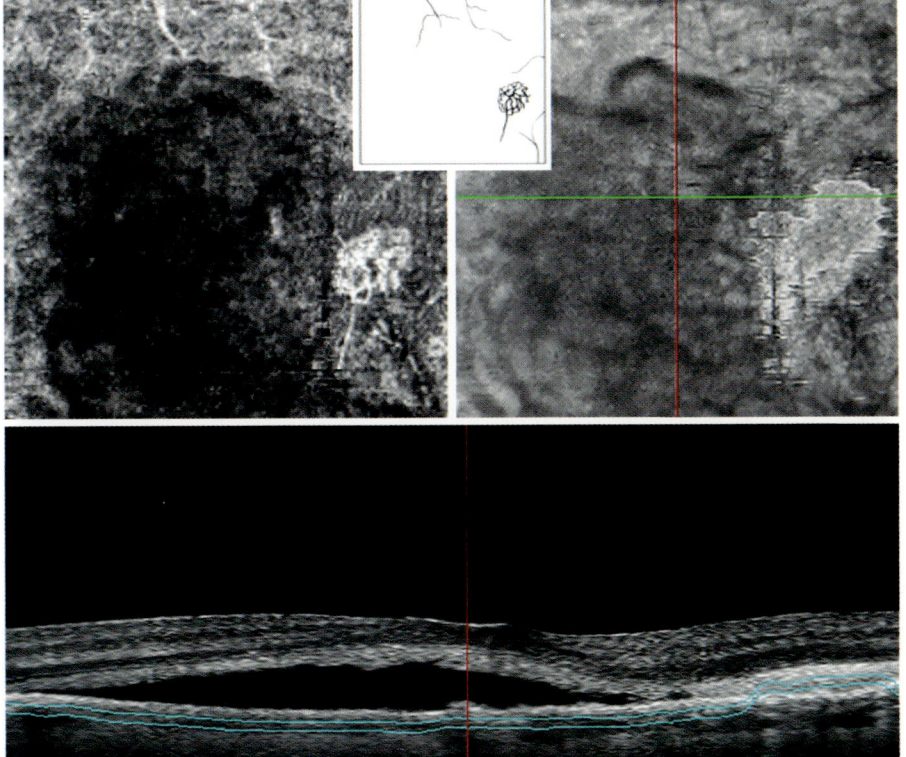

Fig. 7.1.55 Age-related macular degeneration (AMD): Subretinal neovascular membrane of the glomerular type with tangled and twisted new formed vessels. They give the impression of being contained inside a capsule that however is not visible

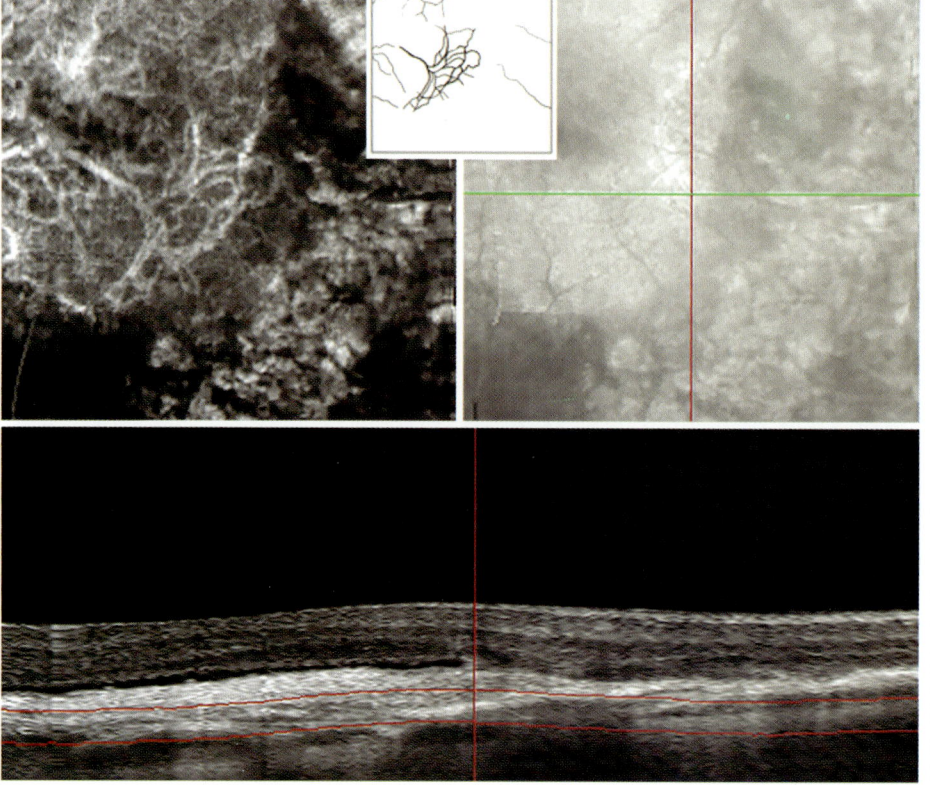

Fig. 7.1.56 Age-related macular degeneration (AMD). Not active neovascular membrane in macular degeneration treated ten times with anti-vascular endothelial growth factor (VEGF). We can see residual sparse subretinal vessels with evident blood flow. Fan-like pattern with feeder vessel. The finer branches are not visible (perhaps occluded) and in any case there is no blood flow in the finer capillaries

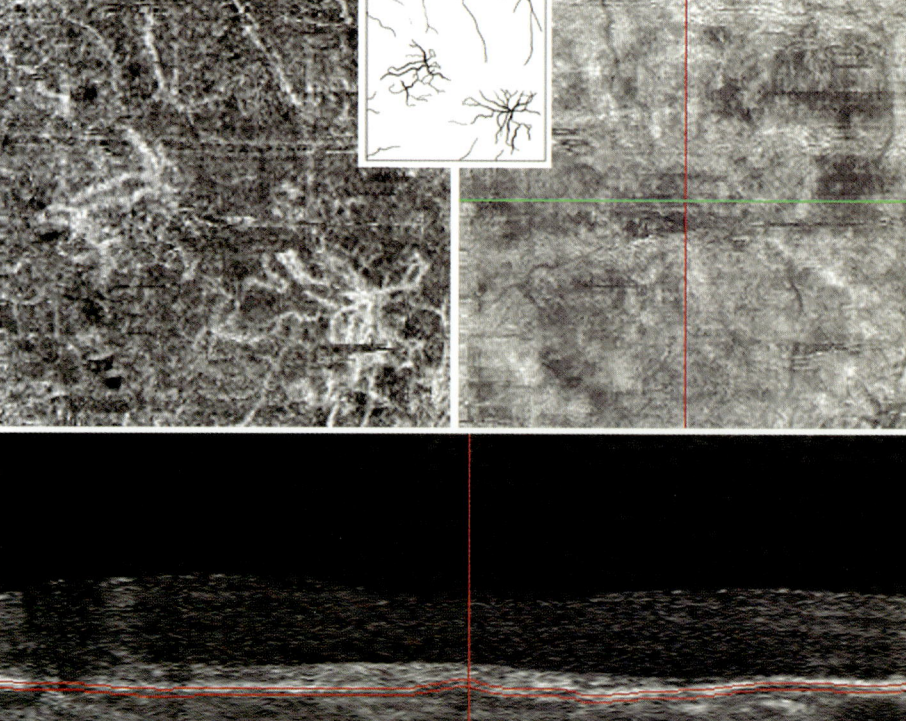

Fig. 7.1.57 Age-related macular degeneration (AMD). Two inactive neovascular membranes in macular degeneration treated several times with anti-vascular endothelial growth factor (VEGF). Coarse networks with blood flow are visible that form an irregular spiderweb. The feeder vessels are visible. The thinner branches cannot be seen (perhaps occluded). No blood flow

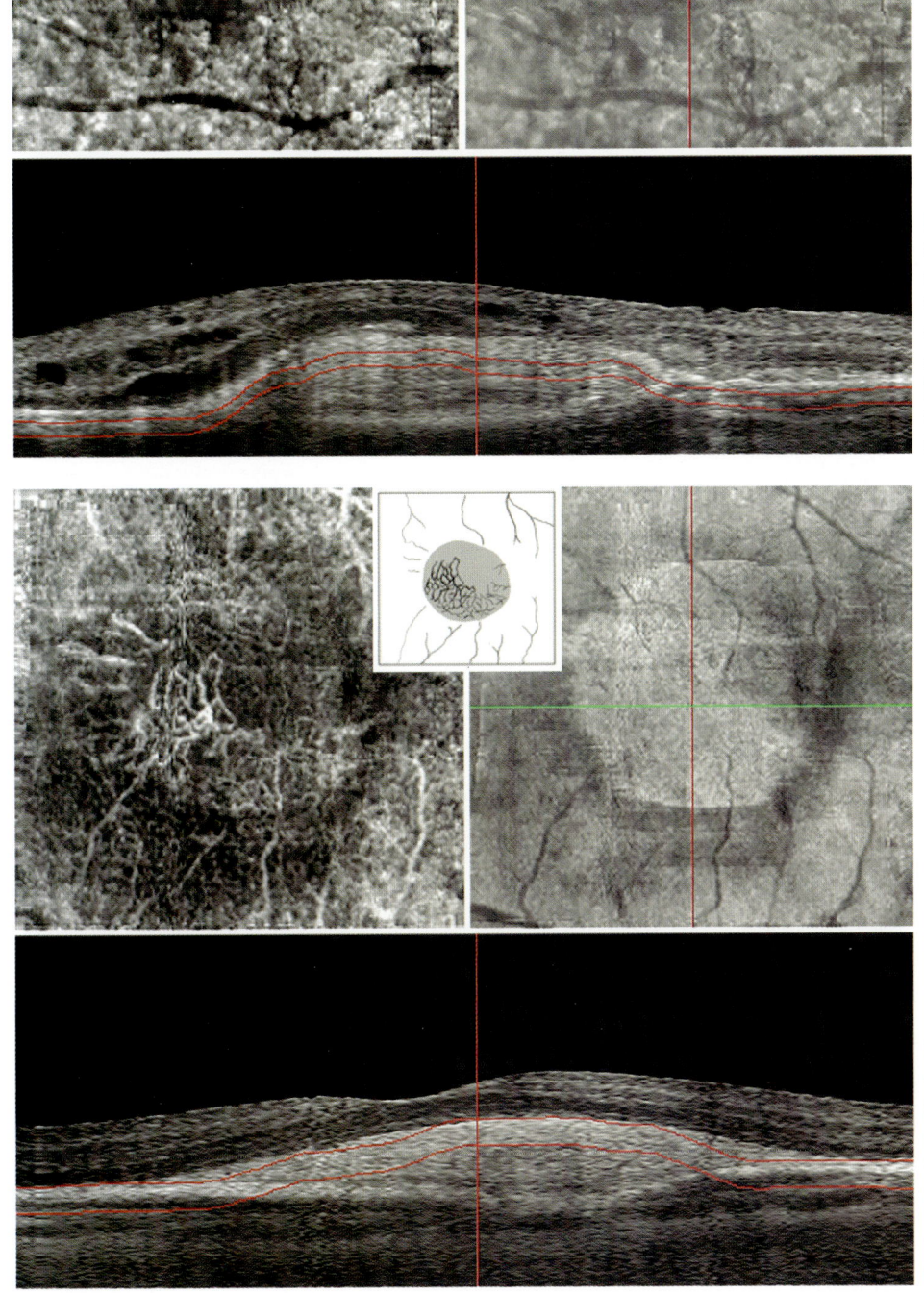

Fig. 7.1.58 Age-related macular degeneration (AMD), macular degeneration with residual fibrovascular membrane. A network of fairly irregular important vessels can be observed while thinner capillaries are not visible because they have no blood flow or are thrombosed. The residual neovascular membrane is contained within an oval-shaped fibrous formation

Fig. 7.1.59 Age-related macular degeneration (AMD): Apparently inactive, long-standing fibrovascular membrane. Inside a rounded fibrous formation, an irregular capillaries network with blood flow. Thinner capillaries are not visible. They could be obliterated or thrombotic. Even though the lesion is not active there are still thin vessels or channels with circulating blood

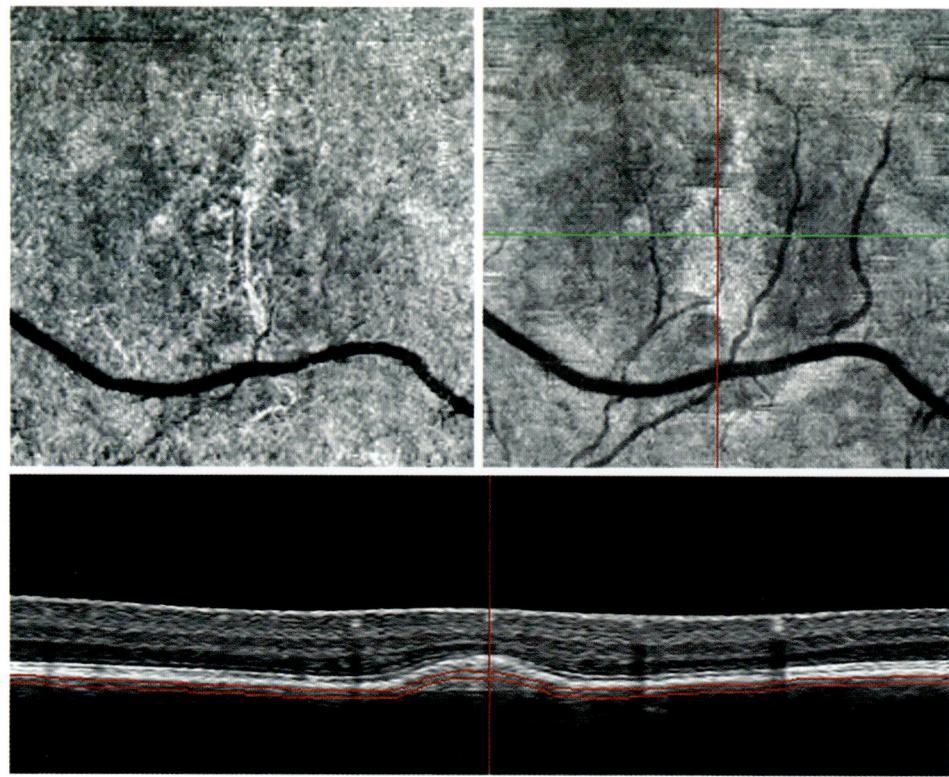

Fig. 7.1.60 Choroidal break scar with inactive fibrovascular membrane; under the pigment epithelium some dilated neovascular capillaries are visible. The thinner capillaries cannot be seen because they are occluded or thrombotic. Even though the lesion is not active, we can still see thin vessels with blood flow

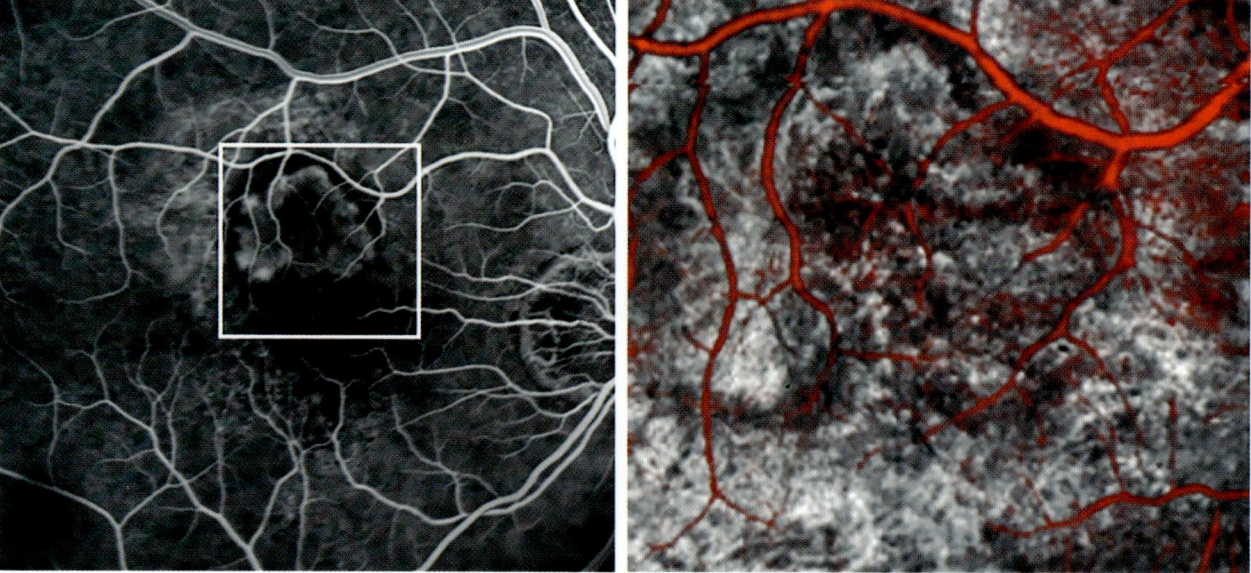

Fig. 7.1.61 Age-related macular degeneration (AMD): Active neovascular membrane. Compare with the fluorangiography with dye leakage. In optical coherence tomography-angiography (Angio-OCT), the vascular branches are sharper. The neovascular membrane is observed inside a fibrovascular formation. Vessels with blood flow form a fragmented and dense, irregular network entirely contained inside the fibrosis margins
Courtesy: Luca Di Antonio, Chieti Pescara

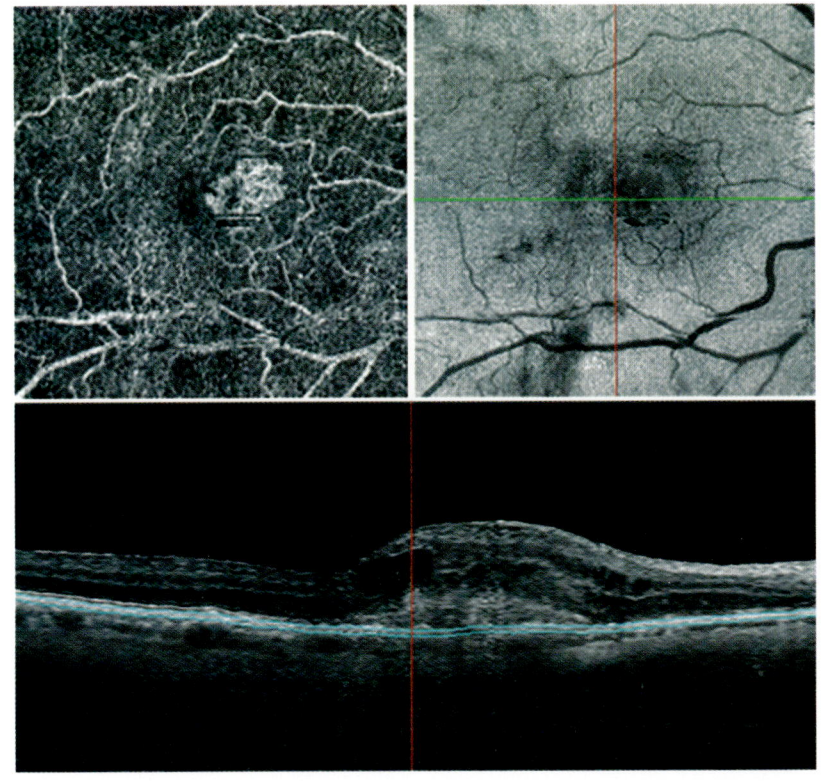

Fig. 7.1.62 Age-related macular degeneration (AMD): Neovascular nodular subretinal membranes. New vessels are tangled and intricate. They seem limited inside a capsule that however cannot be seen
Courtesy: Adil El Maftouhi, Paris

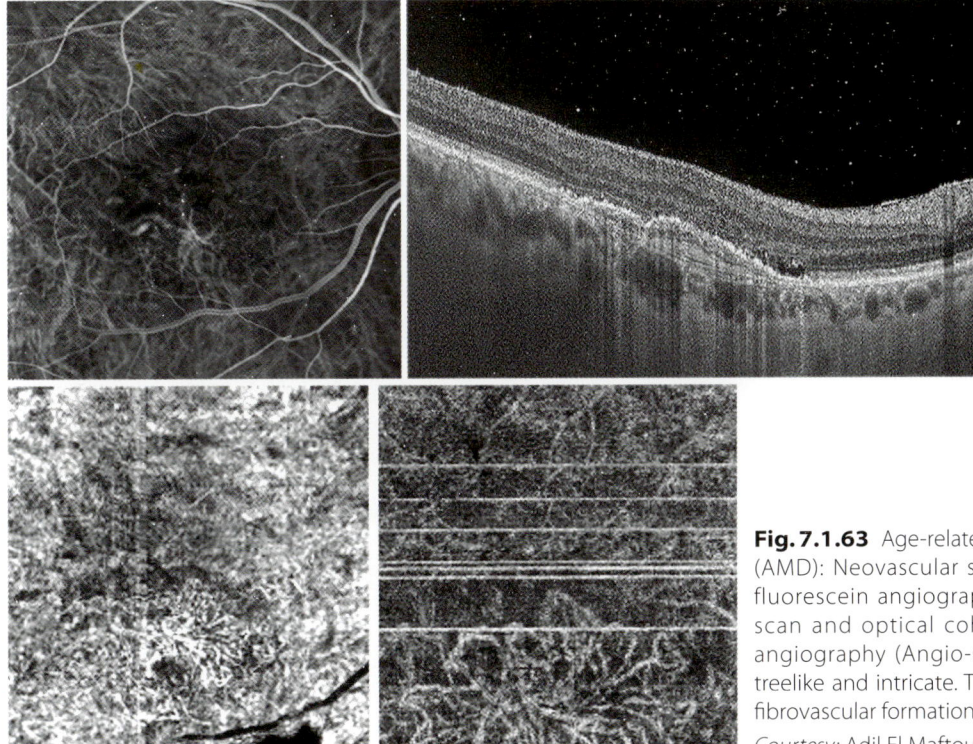

Fig. 7.1.63 Age-related macular degeneration (AMD): Neovascular subretinal membranes; fluorescein angiography (FA), cross-section scan and optical coherence tomography-angiography (Angio-OCT). New vessels are treelike and intricate. They are located inside a fibrovascular formation
Courtesy: Adil El Maftouhi, Paris

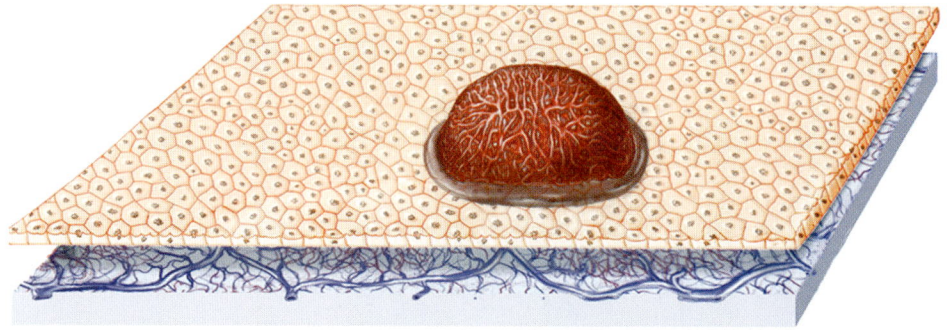

Fig. 7.1.64 Myopic neovascular membrane: Globular feature, with convoluted, tangled, intertwined new vessels forming a globular mass with sharp margins. Around the membrane a pigmented ring can be seen

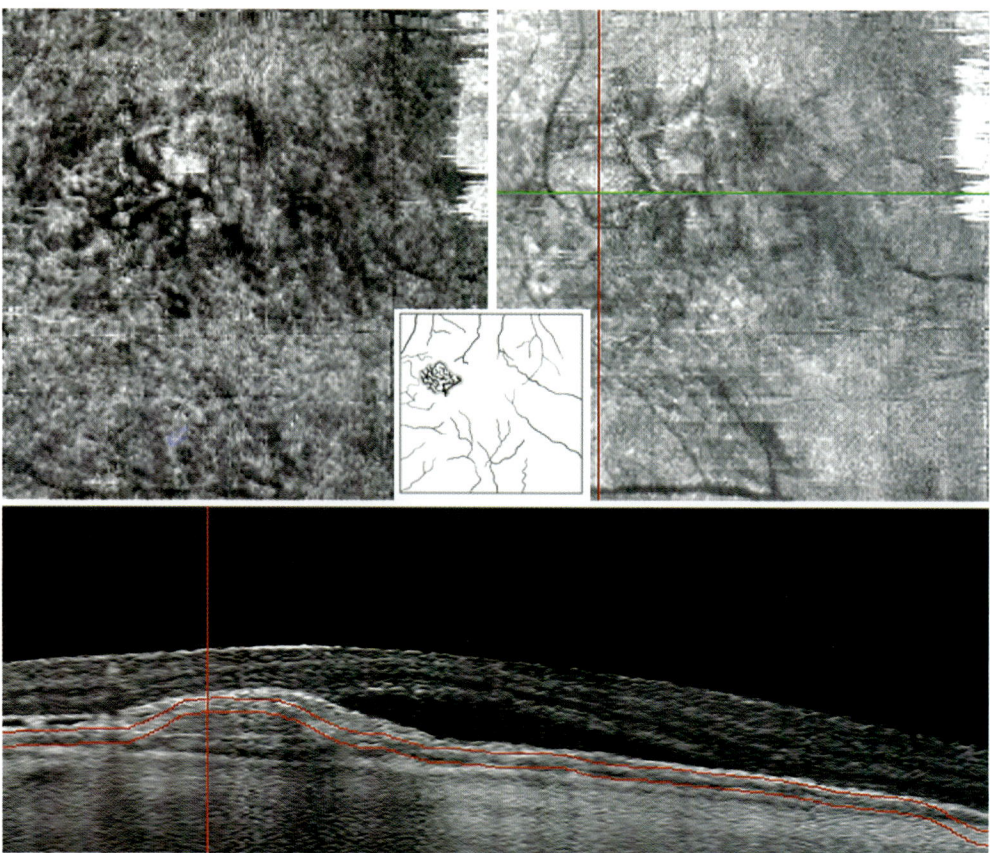

Fig. 7.1.65 Myopic neovascular membrane: glomerular aspect, i.e. globular, intricate, tangled, intertwined forming a globular mass with sharp borders and apparently surrounded by thin capsule

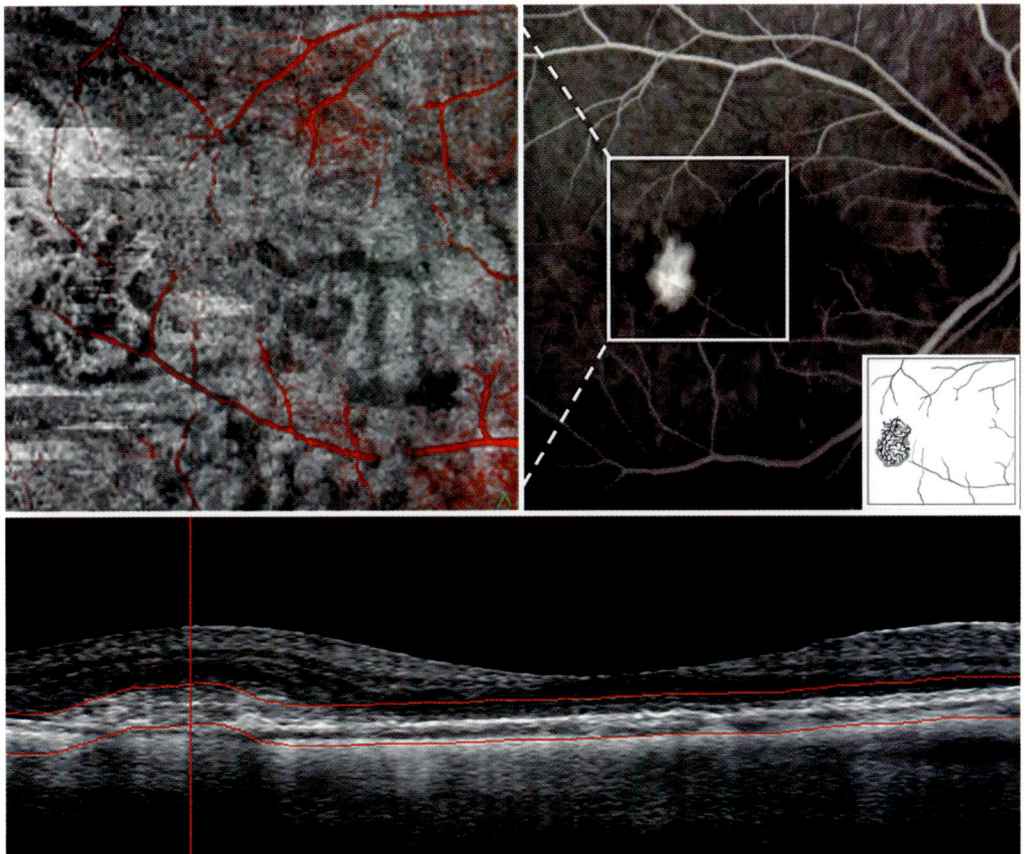

Fig. 7.1.66 Myopic neovascular membrane: The membrane (Left) Angio-OCT is large in size with sharp borders are sharp, and perhaps surrounded by thin capsule. The newly formed capillaries are intertwined, convoluted with a dense capillary network. (Right) Fluorangiography of the same case where dye leakage prevents visualization of the membrane vascular network

Courtesy: Luca Di Antonio, Chieti Pescara

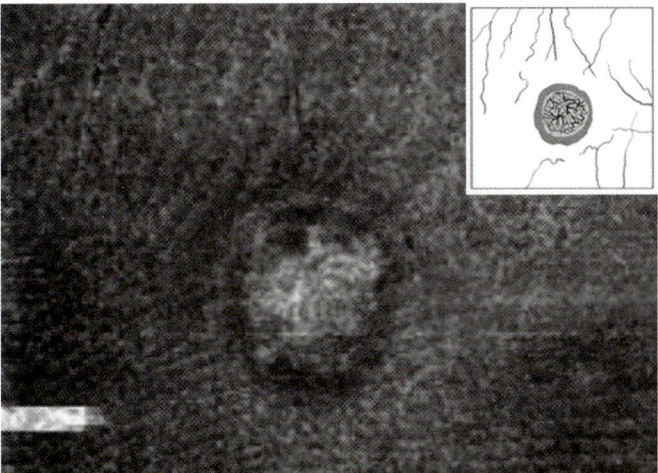

Fig. 7.1.67 Myopic neovascular membrane: globular aspect, intricate, convoluted, tangled, intertwined with sharp borders and apparently surrounded by thin capsule

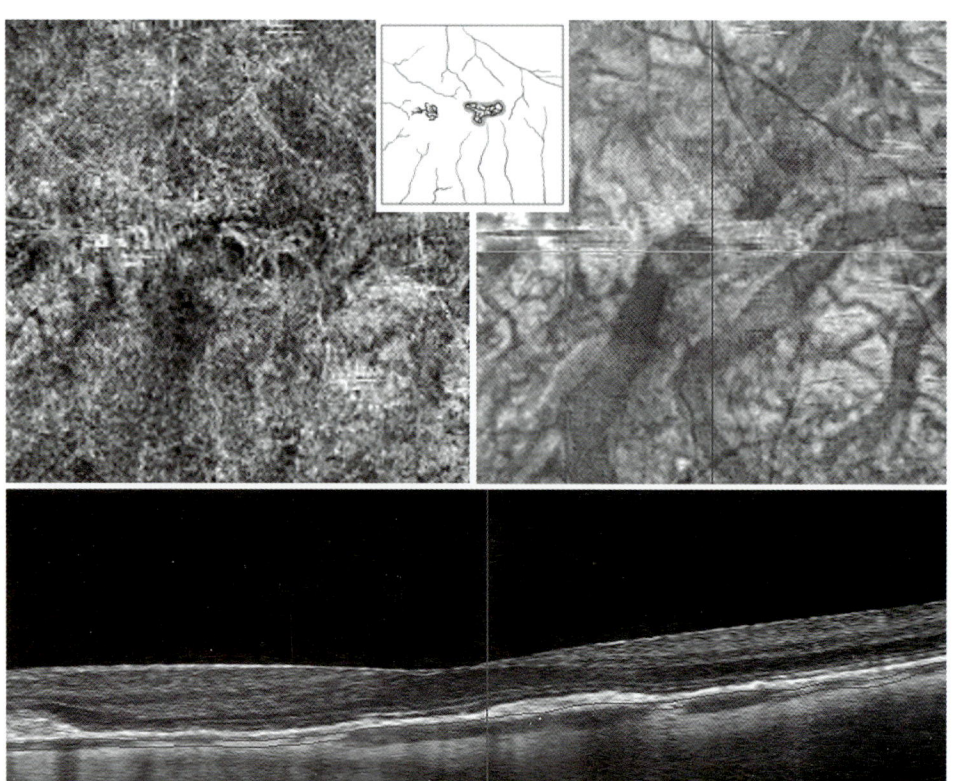

Fig. 7.1.68 Myopic neovascular membrane developed at a longstanding Bruch's membrane (BM) break. We see a tangled irregular, not very dense, vascular network

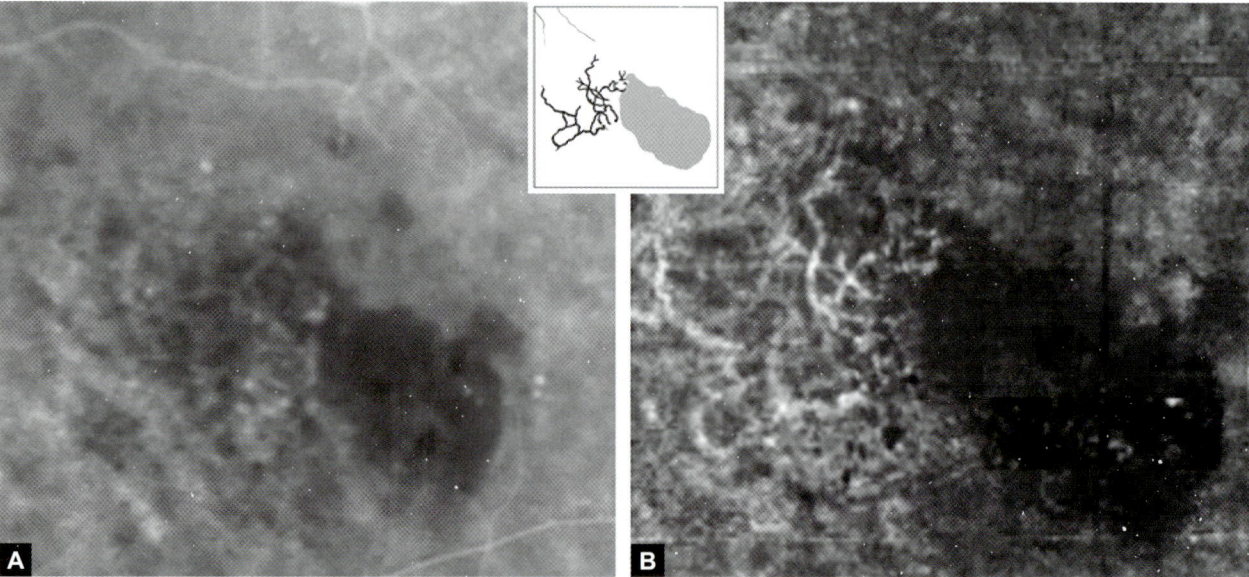

Figs 7.1.69A and B Idiopathic polypoidal choroidal vasculopathy (IPCV). (A) Fluorangiography showing roundish hyperfluorescent formations close to dark patches (screen effect, hemorrhages); (B) Optical coherence tomography-angiography (Angio-OCT) showing a vascular network that interconnects angiomatous formations
Courtesy: Luca di Antonio, Chieti Pescara

GEOGRAPHIC OR ATROPHIC MACULAR DEGENERATION

The incidence of geographic atrophy is much higher than that of the neovascular AMD (Fig. 7.1.70). OCT detects reduction in retinal thickness and increased laser light penetration in the choroid. There is a sharp border between atrophic retina and normal retina. OCT determines the extent of the atrophic area and monitors its evolution. Choroid is thinner than normal at the age of the patient. Sattler and Haller layers are involved in the atrophic process. The vessel walls and connective interstitial tissue between vessels appear denser and flow is evident. Vessel diameter is decreased.

OPTIC DISK DISORDERS

Healthy optic disks show normally a dense vascular capillary network. In case of optic atrophy, glaucoma and optic neuritis, the capillary network may present typical anomalies (see Chapter 8). In case of optic disk atrophy, the vascular network is quite sparse in some points, with loss of the smaller vascular branches (Figs 7.1.71A and B).

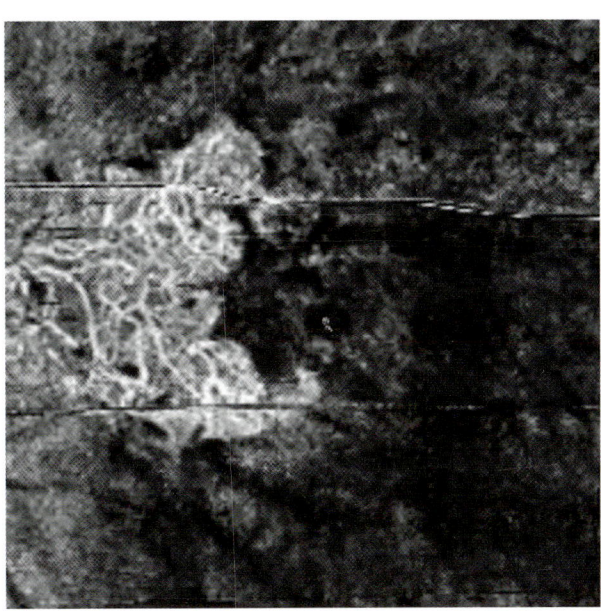

Fig. 7.1.70 Age-related macular degeneration (AMD). Geographic atrophy: In the pigment epithelium and choroidal atrophic area, flowing sclerotic vessels can be seen. They form an irregular network inside an atrophic area with window effect
Courtesy: Adil El Maftouhi, Paris

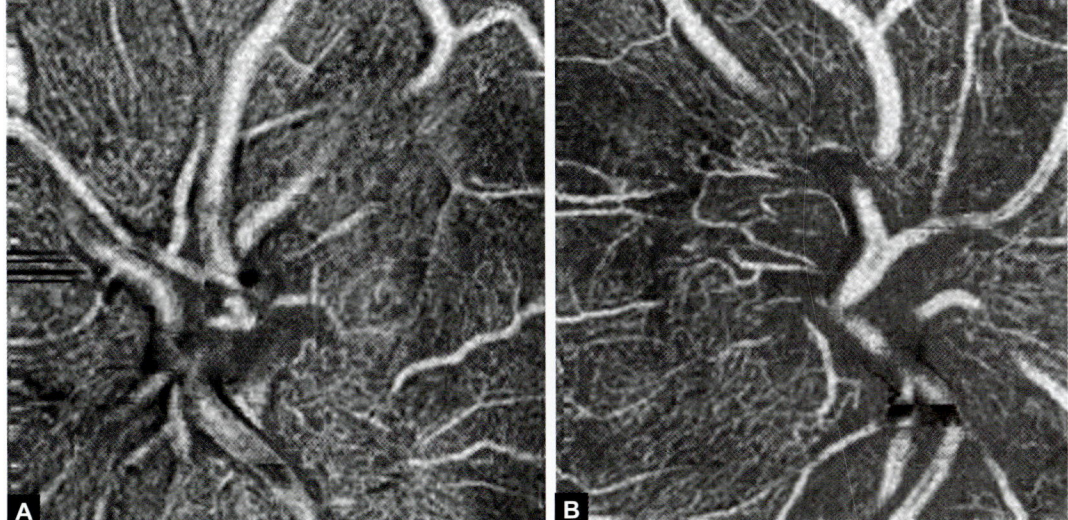

Figs 7.1.71A and B (A) Normal optic disk; (B) A case of optic atrophy showing a sparse capillary network with loss of smaller branches

CHAPTER
7.2

Clinical Applications: Optical Coherence Tomography-Angiography of Choroidal Neovascularization in Age-Related Macular Degeneration

Yali Jia, David Huang

INTRODUCTION

Neovascular age-related macular degeneration (AMD) is an advanced form of macular degeneration. It is characterized by the presence of choroidal neovascularization (CNV). To diagnose neovascular AMD and evaluate the efficacy of treatment, determination of the presence and precise location of the CNV lesion is essential. Currently, structural optical coherence tomography (OCT) has been used to detect small changes in the morphology or the retinal layers and CNV activity, but it cannot reliably discriminate CNV vascular tissue from the surrounding tissue. Our group recently developed a new three-dimensional (3D) ocular angiography using OCT. The algorithm is called split-spectrum amplitude-decorrelation angiography (SSADA).

We used optical coherence tomography-angiography (Angio-OCT) with SSADA algorithm to investigate CNV associated with neovascular AMD. To enhance visualization of CNV, the 3D SSADA angiogram was separately projected into en-face views (Figs 7.2.1A1 to D3) in three layers using an automated algorithm. The inner retinal layer was defined from the internal limiting membrane (ILM) to the outer boundary of the outer plexiform layer (OPL). Thus defined, the inner retina should contain all of the normal retinal vasculature. The outer retinal layer was defined from the OPL to the Bruch's membrane (BM). Because the outer retina is normally avascular, any flow in this layer could be interpreted as CNV. The choroidal layer was defined as below BM. All of these boundaries were identified through the analysis of the reflectance and reflectance-gradient profiles in depth. Clinician's interpretation and manual identification of BM and OPL was necessary when pathologies, such as pigment epithelial detachment (PED) and intraretinal fluid obscured the outer retinal landmarks (AMD case in Figs 7.2.2A to I). Separate en-face images of the inner retina, outer retina and choroid were presented in a sepia color scale. By overlaying color-coded SSADA and structural OCT, the cross-sections (Figs 7.2.1A1 to A3) demonstrated the CNV positions relative to retina pigment epithelium (RPE), which helps to define the types of CNV. Figures 70A1 to D3 show a representative age-matched normal control case (A1 to D1), a type I CNV under subretinal hemorrhage (Figs 7.2.1A2 to D2), a type II CNV (Figs 7.2.1A3 to D3). Figures 7.2.2A to I show a combined type CNV that is hidden on fluorescein angiography (FA) but visible on SSADA. All the cases shown in this chapter were acquired with 100 kHz 1050-nm swept-source OCT.

Furthermore, the blood flow and vessel density of CNV can be quantified.[1] To quantify the blood flow within the CNV, the CNV area and flow index were calculated from the two-dimensional maximum projection outer retina CNV angiogram. The CNV area was calculated by multiplying the number of pixels (for which the decorrelation value was above that of the background) and the pixel size. The CNV flow index was the average decorrelation value in the CNV region.

Optical coherence tomography-angiography are able to better visualize CNV by color coding and quantify CNV blood flow and its activity; however, there are several limitations compared to conventional FA.

First, the view field is small under current commercial available system, so a larger field will require higher speed

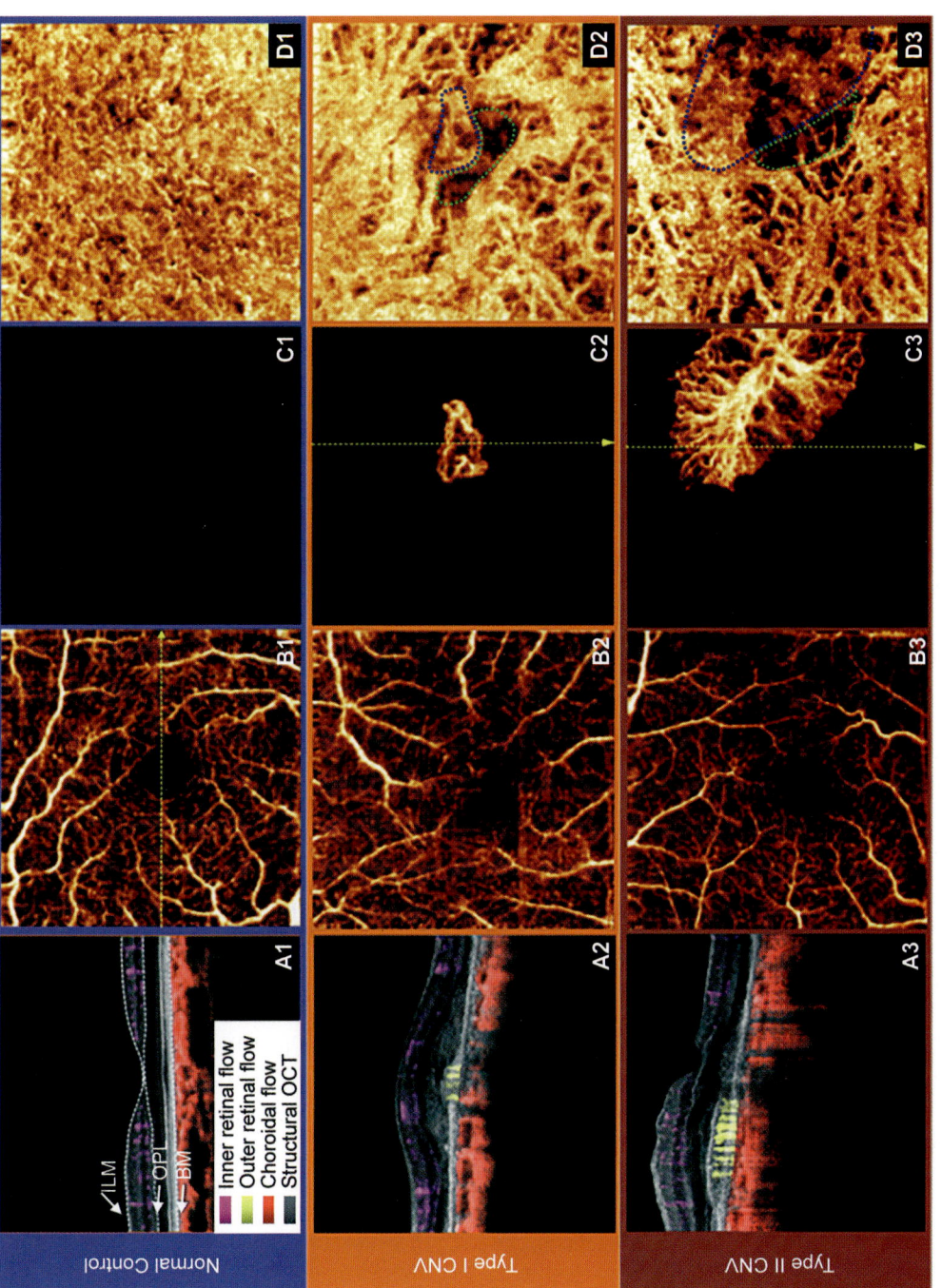

Figs 7.2.1A1 to D3 Cross-sectional color-coded optical coherence tomography (OCT) angiograms (A1 to A3, first column) and en-face OCT angiograms of the inner retina (B1 to B3, second column), outer retina/choroidal neovascularization (CNV) (C1 to C3, third column) and choroid (D1 to D3, fourth column) in normal control (A1 to D1, upper panel), type I CNV under hemorrhage (A2 to D2, middle panel) and type II CNV (A3 to D3, lower panel). The yellow dashed lines indicate the position of OCT cross-section shown in the first column. Shown in A1, the internal limiting membrane (ILM), OPL and BM are the boundaries separating inner retinal, outer retinal and choroidal circulations. The green dotted outlines show the focal region of reduced choroidal flow adjacent to the CNV. The blue dotted outlines show the pathy flow directly under the CNV

Source: Adapted and reprinted with permission from reference 1

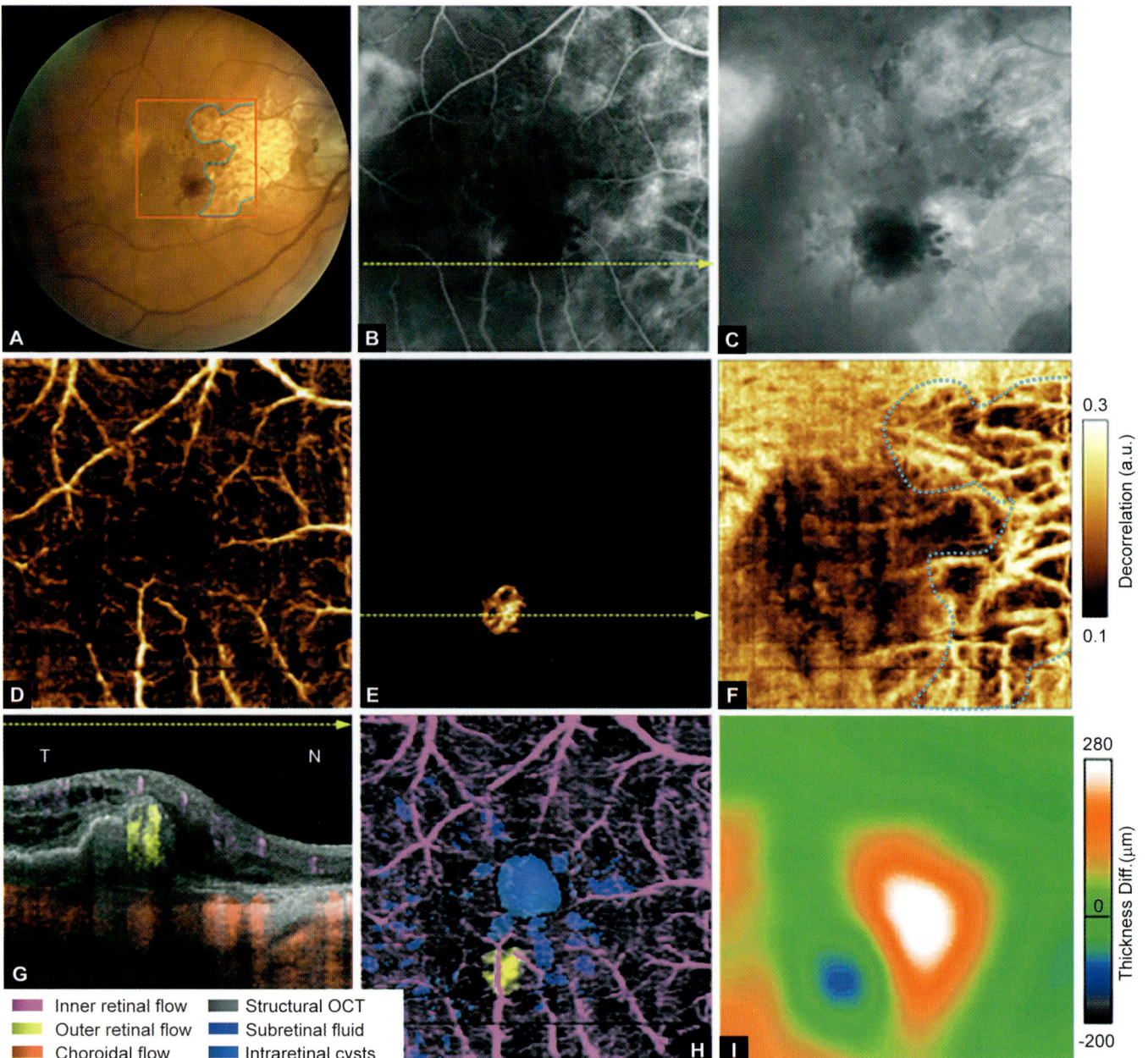

Figs 7.2.2A to I A combined type choroidal neovascularization (CNV) that is hidden on fluorescein angiography (FA) but visible on split-spectrum amplitude-decorrelation angiography (SSADA). (A) Fundus photography showing subretinal hemorrhage, retinal pigment epithelium (RPE) tear and geographic atrophy (blue dashed outline). Red square outlines area shown on angiograms below; (B) Early phase FA; (C) Late phase FA; (D) *En-face* optical coherence tomography (OCT) angiogram of the inner retina; (E) *En-face* angiograms of the outer retina showing the CNV. The yellow dashed lines indicate the position of OCT cross-section shown in panel G; (F) *En-face* angiogram of the choroid showing diffuse reduction of flow signal under the pigment epithelial detachment (PED) and choriocapillaris defect in the area of geographic atrophy (blue dashed outline); (G) Cross-sectional color OCT angiogram showing the CNV both above and below the RPE. The subretinal hemorrhage was over the CNV and overshadowed the CNV at its nasal edge; (H) Composite en-face OCT angiograms; (I) Retinal thickness deviation map showing thinning over the CNV and thickening around it

Source: Adapted and reprinted with permission from reference 1

system. Second, it cannot provide dynamic information, such as transit time and changes in patterns of fluorescence. However, it is possible in the future that combining information from structural OCT, such as fluid, segmentation of RPE and Angio-OCT information, may allow for more comprehensive clinical picture.

REFERENCE

1. Jia Y, Bailey ST, Wilson DJ, Tan O, Klein ML, Flaxel CJ, et al. Quantitative optical coherence tomography angiography of choroidal neovascularization in age-related macular degeneration. Ophthalmology. 2014;121(7):1435-44.

CHAPTER

8

Optical Coherence Tomography-Angiography of Optic Disk and Peripapillary Retinal Perfusion in Glaucoma

Yali Jia, David Huang

INTRODUCTION

Circumstantial evidence suggests that optic disk ischemia may be a causative factor for optic neuropathy disease, such as glaucoma, either by itself or in conjunction with elevated intraocular pressure (IOP). Optical coherence tomography (OCT) angiography is able to measure regional blood flow near the site of injury, in the optic disk and peripapillary nerve fiber layer (NFL). Injury to the nerve fibers due to high IOP or ischemia could reduce the functioning of the nerve fibers before the structures are lost. And this would be reflected in reduced blood flow in the optic disk and the surrounding NFL.

Quantification of Optic Disk Perfusion with 100 kHz 1,050 nm Swept-source Optical Coherence Tomography (Figs 8.1A to H)

To quantify disk blood flow, cross-sectional reflectance intensity images and flow images were summarized and viewed as an en face maximum projection. The disk boundary for each subject was manually delineated along the neural canal opening using the OCT reflectance images of normal (Fig. 8.1B) and perimetric glaucoma (PG) (Fig. 8.1F) eyes. The boundaries were then transferred to the OCT angiogram (Figs 8.1C and G) for disk region segmentation. The disk flow index was defined as the average decorrelation value within the disk.

In a pilot study[1] (24 normal and 11 glaucoma subjects), using 100 kHz swept-source OCT (SS-OCT), we found a dense microvascular network which was visible on OCT angiography in normal disks. This network was visibly attenuated in subjects with glaucoma. The intravisit repeatability, intervisit reproducibility and normal population variability of the optic disk flow index were 1.2%, 4.2% and 5.0% coefficient of variation (CV), respectively. The disk flow index was reduced by 25% in the glaucoma group (P = 0.003). The flow index was highly correlated with visual field (VF) pattern standard deviation (PSD) (R^2 = 0.752, P = 0.001). These correlations were significant even after accounting for age, cup-to-disc area ratio, NFL and rim area.

Quantification of Peripapillary Retinal Perfusion with 70 kHz 840 nm Spectral Optical Coherence Tomography

Although the 70 kHz RTVue-XR spectral OCT is slower than the 100 kHz SS-OCT prototype, it demonstrates better angiography performance for small vessels than SS-OCT due to its better phase and amplitude stability, enhanced split-spectrum amplitude-decorrelation angiography (SSADA) algorithm and longer time interval between consecutive image frames. However, larger disk vessels sometimes show a loss of signal due to the higher sensitivity of spectral OCT to the fringe washout effect associated with high flow velocity. Thus disk flow quantification is problematic on the RTVue-XR. Therefore, in our pilot study (30 normal and 13 glaucomatous subjects) with RTVue-XR spectral OCT, we developed new algorithm to quantify flow index and vessel density in peripapillary retina.

We found that this new parameter provides excellent differentiation between normal and glaucoma (Figs 8.2A to F). Angiography-based flow index and vessel density of peripapillary NFL were both significantly reduced in a group of glaucoma (80% early stage) and suspect subjects. There was very little overlap between the normal

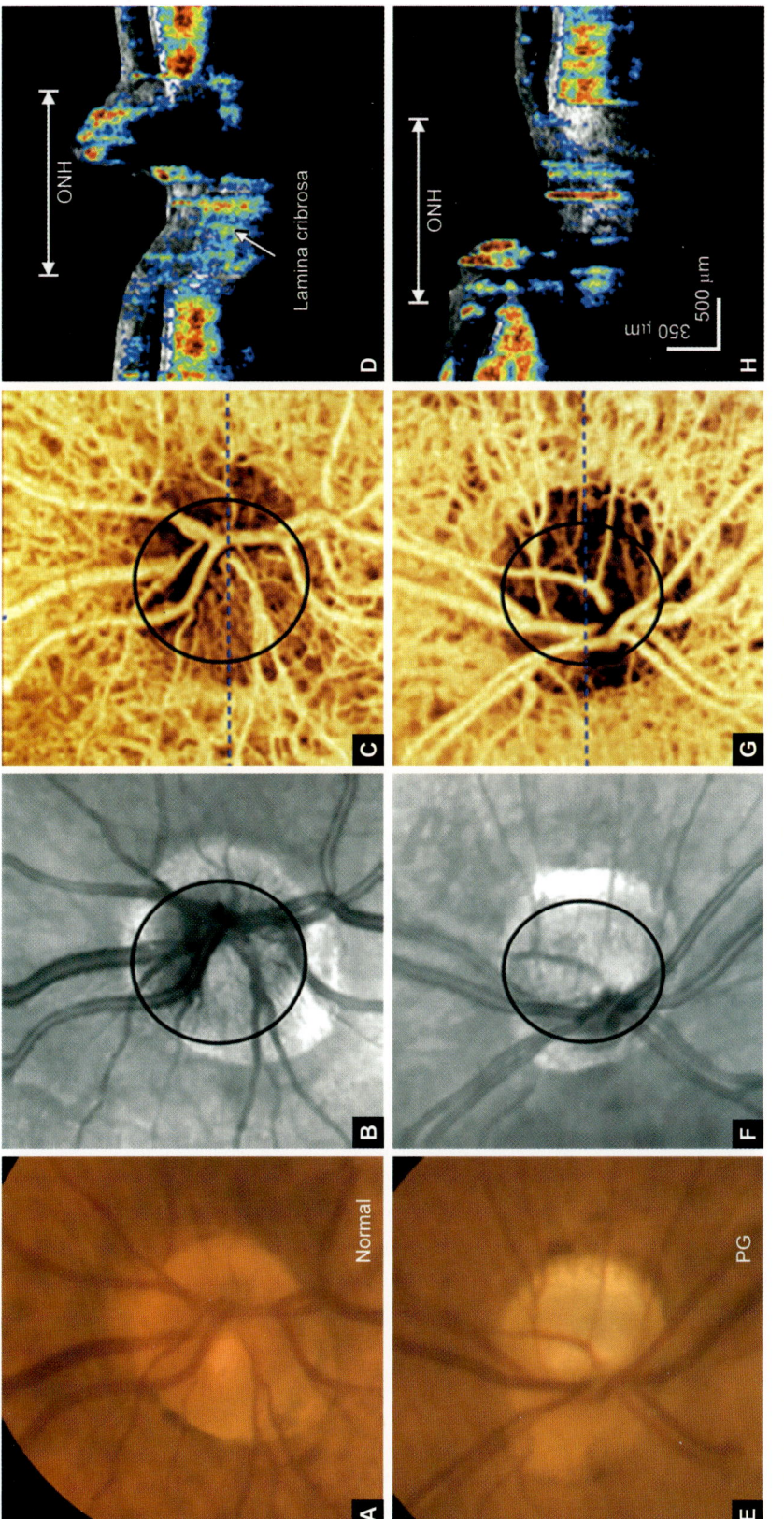

Figs 8.1A to H Disk photographs (A, E), optical coherence tomography (OCT) reflectance (B, F), OCT angiograms (C, G, en-face maximum projection) and cross-sectional angiograms (D, H, overlaying on OCT reflectance in gray scale) in representative normal (A to D) and perimetric glaucoma (PG) subjects (E to H). Images were obtained from the right eye of the normal subject and the left eye of the glaucoma patient. Disk margins are marked by the ellipses. The positions of cross-sections (D, H) are shown by the dashed lines on OCT angiograms (C, G)

Source: Adapted and reprinted with permission from reference 1

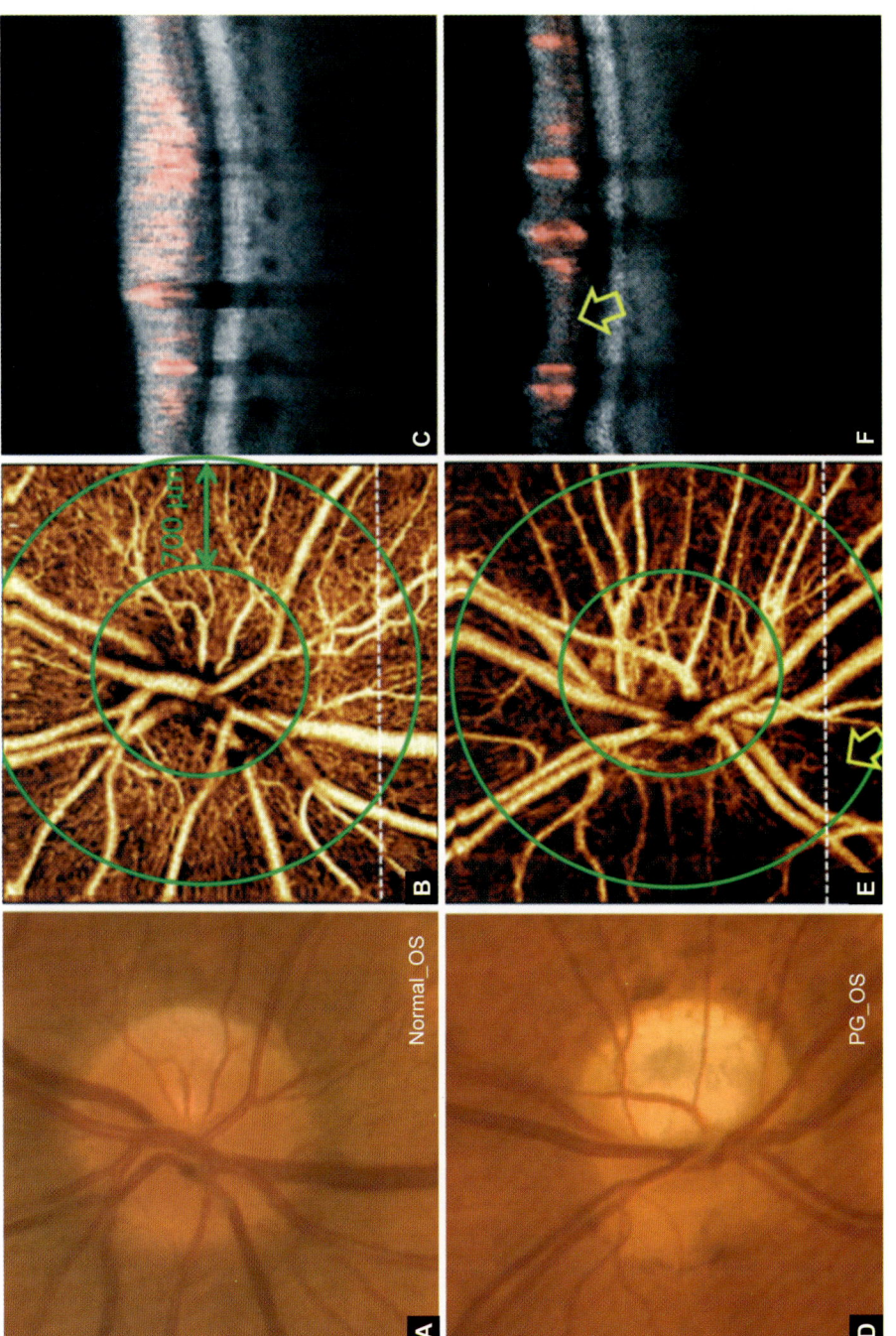

Figs 8.2A to F Disk photographs (A, D), maximum projection of optical coherence tomography (OCT) angiograms (B, E) of disk and retina (3 mm × 3 mm) and cross-sections overlaying retinal blood flow (red) on OCT reflectance (gray) in representative normal (A to C) and perimetric glaucoma (PG) subjects (D to F). Both examples are from left eyes. A dense microvascular network was visible on the OCT angiography of the normal peripapillary retina (the region between two green circles). This network was greatly attenuated and a region of capillary dropout can be clearly identified (yellow arrows) on the en face angiogram of the glaucomatous eye (E). Cross-sectional view (F) showed that the retinal capillary dropout occurred in the area of nerve fiber layer (NFL) thinning. Vessel density and flow index in the annular region around the disk (between green circles in B and D) shows lower density and flow in glaucomatous eyes

and glaucoma groups. The peripapillary NFL circulation highly correlated with VF indices in the glaucoma group and provides high diagnostic accuracy. Furthermore, we have found measurements obtained from OCT angiography to be highly repeatable and reproducible.

REFERENCE

1. Jia Y, Wei E, Wang X, Zhang X, Morrison JC, Parikh M, et al. Optical coherence tomography angiography of optic disc perfusion in glaucoma. Ophthalmology. 2014;121(7):1322-32.

CHAPTER 9

Fluorescein Angiography and Optical Coherence Tomography-Angiography: Advantages and Disadvantages

Bruno Lumbroso, Marco Rispoli

INTRODUCTION

Fluorangiography is currently the most used technique for the vascular network evaluation in research and clinical practice. It is widely used in the study, diagnosis, treatment planning and treatment follow-up in a great variety of ocular disorders, some of which require repeated imaging. Optical coherence tomography (OCT)-angiography (Angio-OCT) is a recent diagnostic technique that presents advantages and disadvantages compared to fluorangiography.

Fluorescein angiography (FA) uses injection of fluorescein dye to obtain the vascular map of the eye. Angio-OCT we use in this book uses split-spectrum amplitude-decorrelation angiography (SSADA) algorithm to enhance all moving structures in a scan, shadowing the not moving structures. OCT acquires five fast B-scans on the same point and measures motion differences.

GENERAL FLUORANGIOGRAPHY ADVANTAGES

- It is a well-known technique, currently the gold standard in retinal angiography.

GENERAL FLUORANGIOGRAPHY DISADVANTAGES

- Fluorescein angiography has light adverse effects as nausea, fainting that are frequent.
- Serious adverse effects are rare but can be dangerous or, rarely, life threatening: allergy, cutaneous rashes, Quincke's edema, infarcts, pulmonary edema, anaphylaxis, shock.
- Adverse effects are more frequent if fluorangiography is frequently repeated. Fluorescein should not be used in pregnancy, in children or in patients with renal or cardiac disorders. In reason of risk of undesirable side effects of fluorescein, this invasive examination is not performed unless a definite clinical condition is known to exist.

FLUORANGIOGRAPHY IMAGING ADVANTAGES

- Fluorangiography clinical use allows visualization of fluorescein leakage, fluorescein pooling in cavities as microaneurysms. Tissue staining, leakage is a most important symptom in diabetic retinopathy and in vascular occlusion from retinal microvascular anomalies, microaneurysms and neovascularization.
- Neovascularization and intravitreal antiangiogenic therapies influence vessel permeability, affecting measurements of lesion size by fluorangiography. Fluorangiography images may show with great precision the capillary net around the foveal avascular zone, in the first frames of the examination.
- Similarly, when the optic nerve head is imaged, the high permeability of the ciliary capillary network to fluorescein may mask vascular details that would be detectable on Angio-OCT.
- Dye leakage, pooling in cavities and tissue staining help greatly diagnosis in vascular and inflammatory diseases.

GENERAL ANGIO-OCT DISADVANTAGES

There is no more discomfort associated to Angio-OCT than with a normal fundus camera.

GENERAL ADVANTAGES OF ANGIO-OCT IMAGING

- It is noninvasive and therefore used to follow evolution of vascular disorders at frequent intervals, without problems for the patient. It can be safely and easily repeated, and allows repeated and frequent follow-up imaging and makes possible screening.

ANGIO-OCT IMAGING ADVANTAGES

- A noninvasive imaging procedure, such as Angio-OCT, is frequently a good alternative to fluorangiography for monitoring, in case of cardiology patients, pregnancy and children.
- Optical coherence tomography-angiography can study the foveal capillary network in its multiple planes in the different layers of the retinal tissue. It visualizes vascular circulation using blood motion contrast and has the advantage of three-dimensional imaging, thus displaying several vascular layers. Using this method, two separate layers of microvasculature are observed in the ganglion cell layer, the anterior and the posterior inner nuclear layers.
- Capillaries in retina and macula and foveal capillary arcade are better observed than in fluorangiography where they can be seen only in the early frames.
- Lack of leakage allows seeing capillary abnormalities and choroidal neovascularization (CNV) net better than with fluorangiography.
- Lack of wall staining allows observing better the blood column abnormalities.
- Optical coherence tomography-angiography provides more vascular information, for example, on arteriovenous shunts, collaterals and dilated blood vessels.
- Whereas Angio-OCT images do not reveal leaks, a critical diagnostic parameter, the technique can, as a consequence, provide an alternative means of assessing lesion size, which may be overestimated by fluorangiography when dye leakage is important.

ANGIO-OCT DISADVANTAGES

- Presently OCT devices allow only a limited field of view (Tables 9.1 and 9.2).
- Optical coherence tomography-angiography images do not show leakage, dye pooling and tissue staining. They allow assessing lesion size, which can seem greater where dye leakage is important. In the ischemic territories, capillaries thin out and retina texture is different.
- Hemorrhages hide slightly the capillaries. Exudates are not seen easily.

Table 9.1 SSADA Angio-OCT differences with FA are immediately evident

- No dye leakage/pooling/staining on the Angio-OCT.
- In the ischemic territories, capillaries thin out and retina texture is different.
- Hemorrhages hide slightly the capillaries.
- Exudates are not always seen.
- Microaneurysms are not all seen, but only the bigger ones.
- When in FA there is leakage of the dye, in Angio-OCT there is a slight blurring of the capillaries.
- Wall staining is not seen in Angio-OCT, but we can see a dark abnormal vessel wall.
- Blood column is very thin.

Abbreviations: SSADA, split-spectrum amplitude-decorrelation angiography; Angio-OCT, optical coherence tomography-angiography; FA, fluorescein angiography

Table 9.2 Some features are better seen with SSADA Angio-OCT

- Capillaries in retina and macula better seen than in FA.
- Foveal capillary arcade better seen than in FA where it is seen only in the early frames.
- Lack of leakage allows seeing capillary abnormalities better than with FA.
- Lack of leakage allows seeing capillary loops and shunts.
- Lack of leakage allows seeing better the CNV net.
- Lack of wall staining allows seeing better blood column abnormalities.
- Lack of leakage staining allows seeing better optic nerve head capillaries in glaucoma and optic nerve disorders.

Abbreviations: SSADA, split-spectrum amplitude-decorrelation angiography; Angio-OCT, optical coherence tomography-angiography; FA, fluorescein angiography; CNV, choroidal neovascularization

- Microaneurysms are not all seen, but only the bigger ones, probably the ones with more leakage in FA. When in FA there is leakage of the dye, Angio-OCT shows a slight blurring of the capillaries.
- Wall staining is not seen in Angio-OCT, but we see abnormal dark vessel wall.
- Optical coherence tomography-angiography provides more vascular information, for example, on arteriovenous shunts, collaterals and dilated blood vessels than fluorangiography.

OTHER NONINVASIVE TECHNIQUES

- *Retinal function imaging (RFI):* This procedure uses high illumination with stroboscopic light to highlight blood cells in capillaries. Patients do not tolerate well this examination, in reason of intense dazzling.
- *Adaptive optics* ophthalmoscope produces excellent retinal imaging, using blood cells for motion contrast, but the very small limited field of view makes its clinical use difficult.

CHAPTER

10

Reporting an Optical Coherence Tomography-Angiography

Marco Rispoli

INTRODUCTION

Reporting an optical coherence tomography-angiography (Angio-OCT) is strictly linked to en-face OCT interpretation. Angio-OCT and flow indexes are extracted from en-face sections by split-spectrum amplitude-decorrelation angiography (SSADA) algorithm. Angio-OCT shows retinal vasculature layer-by-layer while traditional angiography gives a full thickness view. After segmenting the macular cube and recognizing the vascular structures it is possible to proceed analytically from vitreous towards choroid. Taking this basic principle into account, the reading of an Angio-OCT must be made into two steps:
1. *Analytic steps:* In analyzing an Angio-OCT, the following features must be examined in sequence:
   Localizing the scan depth:
   Flow and decorrelation
   Vascular morphology and architecture
   Texture.
   Followed by:
2. *Synthesis and diagnosis.*

ANALYTIC STEPS

Localizing the Scan Depth

There are four main steps to follow during the report phase:
1. Superficial vascular plexus
2. Deep vascular plexus
3. Retinal pigment epithelium (RPE) and Bruch's membrane (BM)
4. Choroid.
   Correct interpretation requires examination of en-face images at multiple depth levels and cross-sectional images in order to confirm clinical interpretation of an Angio-OCT image at a specific level. For each case and each level we have to analyze:
- *Flow and decorrelation:* Angio-OCT is generated by motion contrast. Blood cell motion is detected using a decorrelation signal from either intensity or phase. Blood flow can be rapid or slow. Its direction can be transverse or vertical; the vessels can be thin or thick.
- *Vascular morphology and architecture:* The architecture of the vascular network may be regular or irregular and it may be dense, sparse, widened, dilated, rarefied, faint, dense and tangled. The meshes may be sharp, distinct or indistinct. Network may be loose, dilated, narrow, rarefied, faint, dense, compact, tangled, sharp or indistinct.

  We must also assess the shape of the vessels that may be regular or irregular and the density of the capillaries that may be sparse or dense. The cross-section of the capillaries shows up their diameter that may be small, large, regular or irregular.
- *Texture:* Texture is a new concept in description of Angio-OCT that is not used in classical structural OCT or in fluorangiography. The texture is the feature made by the threads of a fabric. It is based on vascular density. It can be loose, coarse, large texture, granulated, fine, faint, speckled, subtle, thin and grayish.

This new terminology is essential to report an Angio-OCT.

CASE REPORTS

We describe three case report examples:
1. Diabetic patient without retinopathy
2. Choroidal neovascularization (CNV)
3. Prepapillary new vessels in diabetic retinopathy.

Case 1: Diabetic Patient without Retinopathy

The main steps to follow during the report phase:
- Superficial vascular plexus
- Deep vascular plexus
- Retinal pigment epithelium and BM.

Diabetic subjects with no ophthalmoscopy and angiography signs of retinopathy show very interesting features if compared with traditional angiography (Figs 10.1A and B).

- *Superficial vascular plexus:* A particularly evident capillary network is suggestive of an incipient retinopathy because these changes appear early. They are due to the increase in the size of some capillaries while others are closed and thus we see a looser network with larger and more sparse meshes. There is an increase in the size of the foveal avascular area that normally is about 500 μm large.
- *Deep vascular plexus*: Evidence of deep vascular plexus alterations. They show important modifications: capillaries become rarefied with increased vascular size. This plexus, when congested, is better identified by a thicker segmentation (60 μm or more, dual of normal value).

Deep microvasculature seems to be increased in size with a less organized and regular capillaries, if compared with a normal control. It is very important to study the connections between the superficial and the deep vascular plexus.

Moving dynamically from the retinal surface towards outer retina layers we observe reduced texture density combined with short and large vertical flows connecting superficial and deep vascular plexus.
- *Retinal pigment epithelium* and *Bruch's membrane:* Normal.

Case 2: Choroidal Neovascularization

The main steps to follow during the report phase:
- Superficial vascular plexus
- Deep vascular plexus
- Retinal pigment epithelium and BM.

We recommend to study both (superficial and deep) vascular retinal plexuses, even if they seem to be normal, before proceed analyzing CNV.

Superficial Vascular Plexus

- Normal morphology and texture of the superficial plexus.

Deep Vascular Plexus

- Normal morphology and texture of the deep plexus.

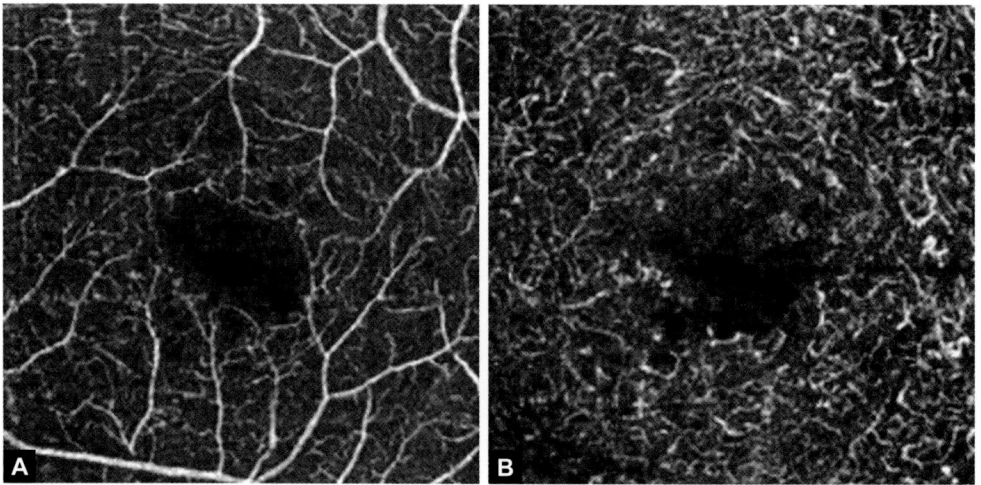

Figs 10.1A and B Diabetic subject without ophthalmoscopic diabetic retinopathy. Superficial vascular plexus appears normal except onset of foveal avascular zone (FAZ) enlargement. Deep vascular plexus alterations are seen clearly. Vessels appear congested with texture fading. Decorrelation seems to be higher than the normal retina

Retinal Pigment Epithelium and Bruch's Membrane

Optical coherence tomography-angiography allows us to locate exactly the CNV in size and depth, since it has neither leakage nor pooling effect, typically from traditional angiography. Outer nuclear layer (ONL) is avascular, so the identification of blood flow inside this layer is strongly evocative of CNV.

An important advice while analyzing a CNV is to verify that we are observing a real neovascular flow and not a "mirror effect" due to SSADA projection of superficial plexus on the RPE. Sometimes it is easy to confuse CNVs with atypical surface retinal vessels. In order to avoid this "effect" it will be sufficient to shift dynamically the CNV segmentation a little up and down, seeing the "mirror effect" disappearing.

It is very important to select the appropriate segmentation thickness while CNVs have a large depth and size variability (subretinal or sub-RPE) that may change during the pathology evolution.

In the age-related macular degeneration (AMD) (Fig. 10.2), CNV usually are located below RPE (type 1) and less frequently above RPE in the subretinal space (type 2). Studying Angio-OCT we talk about blood flows, and only indirectly about vessels.

As a main boundary we will choose the BM because its profile follows the ideal posterior pole concavity (RPEref), and is crossed by CNVs.

In the naïve subjects, CNV looks like a rounded, bicycle wheel network with a feeder flow. Some peripheral anastomosis is also observed. Anastomosis is important during the anti-vascular endothelial growth factor (VEGF) follow-up because their density is reduced after treatment. CNV flows are very clearly seen and it is possible to follow their course and main branches.

After treatment the network is reduced in size, with less anastomoses and important flow modifications. The network flow shows "fragmentation".

In the case presented here, CNV has been treated by 12 ranibizumab injections. The network is reduced in density and in size. There are few anastomosis and important flow modifications. CNV network flow appears "fragmented", octopus-like, and it is difficult to follow the flow course.

Case 3: Prepapillary New Vessels in Diabetic Retinopathy

This young pregnant (4th month) woman presented a sudden visual acuity decrease due to vitreous hemorrhage.

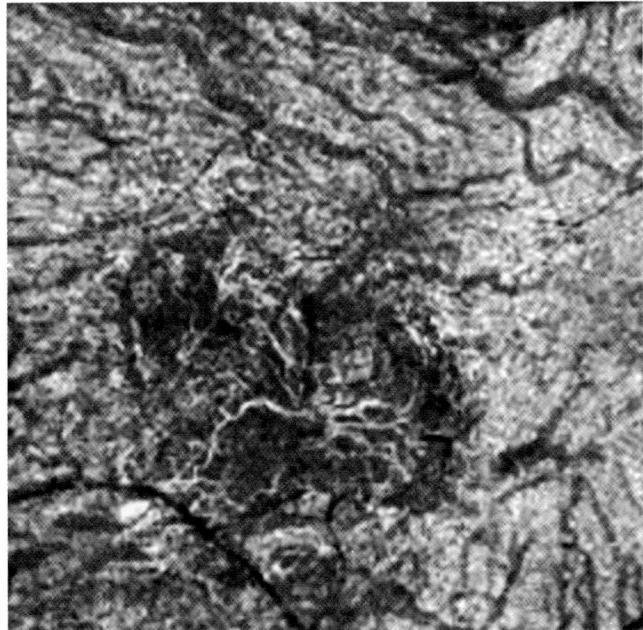

Fig. 10.2 Age-related macular degeneration (AMD). Relapsing neovascular membrane after treatment with anti-vascular endothelial growth factor (VEGF). The neovascular membrane is located inside a fibrovascular formation. The vessels form an irregular network contained inside a fibrotic area
Courtesy: Luca Di Antonio, Chieti Pescara

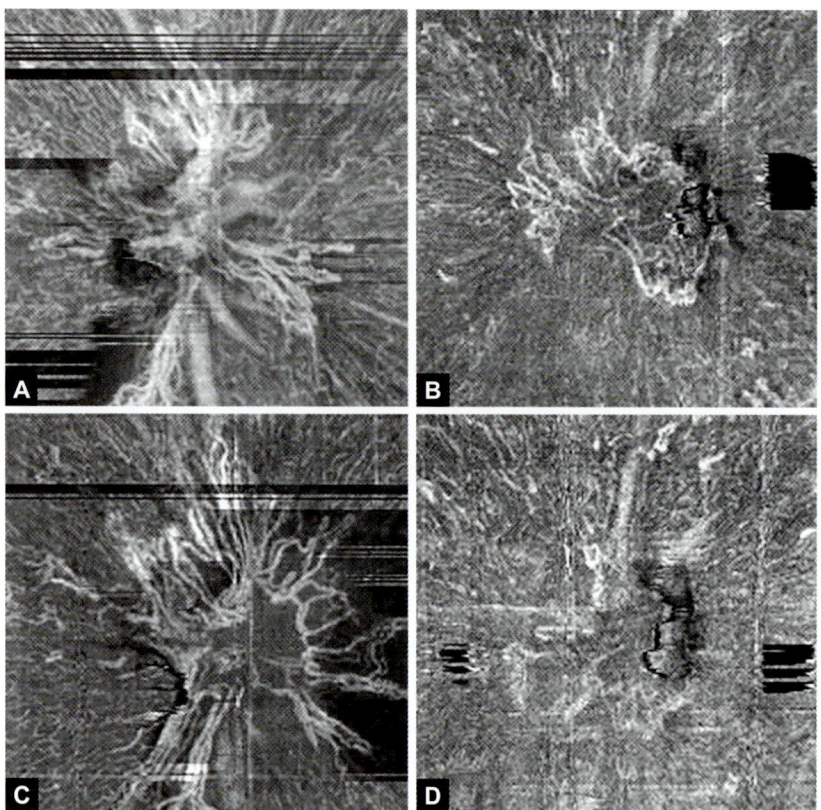

Figs 10.3A to D Proliferative diabetic retinopathy with prepapillary new formed vessels. The patient is a young woman in her 4th month of pregnancy. Fluorangiography could not be performed. Angio OCT is 200 μm thick and the flows were acquired between retinal surface and prepapillary space (B and A). Panretinal photocoagulation was used, and 3 weeks late Angio OCT was repeated using the same scan protocol (C and D) was performed. OCTA shows a flow decrease and a clear reduction of the vessels width

Angio-OCT detected prepapillary and papillary new vessels (Figs 10.3A to D). During diabetic retinopathy, Angio-OCT is able to follow the new vessels regression after laser or anti-VEGF/steroid treatment.

Detecting prepapillary new vessels need a particular procedure. Prepapillary new vessels are detectable by an inner limiting membrane (ILM) thick segmentation slice (250 μm and more). This way of segmentation allows us to identify the new vessels located in the optic disk depression, on the optic disk surface and in vitreal prepapillary space.

Those new vessels are very thin, fanlike, with high flow signal.

After treatment, those new vessels show fragmentation with massive reduction of volume. Branching is rarefied.

CHAPTER 11

Future Ultrahigh Speed Swept-Source OCT Technology and OCT-Angiography

Nadia K Waheed, Woo Jhon Choi, Jay S Duker, James G Fujimoto

INTRODUCTION

This chapter presents an overview of future optical coherence tomography (OCT) technology, focusing on ultrahigh speed, swept-source OCT (SS-OCT) and its application to optical coherence tomography-angiography (OCTA). The technology described here is currently used in research and not yet commercially available, but these methods and results suggest the future potential of OCTA. Imaging speed is especially important because angiographic protocols require repeated scanning of the same position on the fundus in order to generate motion contrast from blood flow. In addition, OCTA is typically displayed *en face*, and therefore each pixel in the image requires multiple axial scans. For this reason, *en-face* OCTA requires significant trade-offs between image acquisition time, retinal coverage and *en-face* pixel resolution. Increasing imaging speed is critical in order to improve retinal coverage and pixel resolution, while still maintaining clinically practical imaging times.

SWEPT-SOURCE OPTICAL COHERENCE TOMOGRAPHY FOR ULTRAHIGH SPEEDS

The most promising approach for achieving ultrahigh speed is a technique known as SS-/Fourier domain OCT (FD-OCT). SS-OCT uses an interferometer with a frequency swept, narrow bandwidth laser, rather than a broadband light source, spectrometer and line-scan camera as in spectral-domain OCT (SD-OCT).[1,2] SS-OCT can achieve much faster imaging speeds than SD-OCT because the speeds are not limited by camera reading rates. In addition, SS-OCT enables imaging with longer wavelength light, at approximately 1,050 nm, a wavelength for which conventional cameras have limited sensitivity. OCT at 1,050-nm wavelength suffers less scattering loss from cataracts or ocular opacities, and tissue image penetration is improved compared to 850 nm typically used in SD-OCT instruments.[3,4] These combined advantages enable SS-OCT to achieve ultrahigh speeds as well as to better image structures, such as the choroid and choriocapillaris.

> **How Swept-source Optical Coherence Tomography Works**
> Figure 11.1 shows a schematic view of how SS-/FD-OCT detection works. SS-/FD-OCT detection uses a narrow bandwidth, frequency swept light source with an interferometer. The interference output is detected with a high speed photodetector, without the need for a spectrometer and line-scan camera. In SS-/FD-OCT detection, different echo time delays of light are encoded as different oscillation frequencies when the laser light source is frequency swept. The echo time delays can be measured by Fourier transforming the detector signal to extract the oscillation frequencies. SS-/FD-OCT detection can also be understood by noting that the light source frequency sweep essentially labels different times with different frequencies. The output from the swept light source is split into two paths. One light path is directed onto tissue, and light is backreflected or backscattered from tissue structures at different depths. The second path is reflected from a fixed (not scanned) reference mirror at a given delay. The light from the tissue and the reference has a relative time delay Δz related to the depth of the tissue structure. The interference of the signal and reference light will produce an oscillation or beat frequency because there is a frequency difference between the two light waves at the detector. The oscillation frequency will be related to the echo time delay Δz. Larger echo delays will produce higher frequency oscillations. Similar to SD-/FD-OCT detection, the echo delays or axial scans can be measured by Fourier transforming the detector signal acquired over one frequency sweep of the light source. Each frequency sweep of the light source generates one axial scan, and the axial scan imaging rate is determined by the sweep repetition rate of the light source.

Swept-source techniques were used 20 years ago for fiber optics/photonics measurement as well as laser radar. SS-OCT was demonstrated by our group as early as 1997

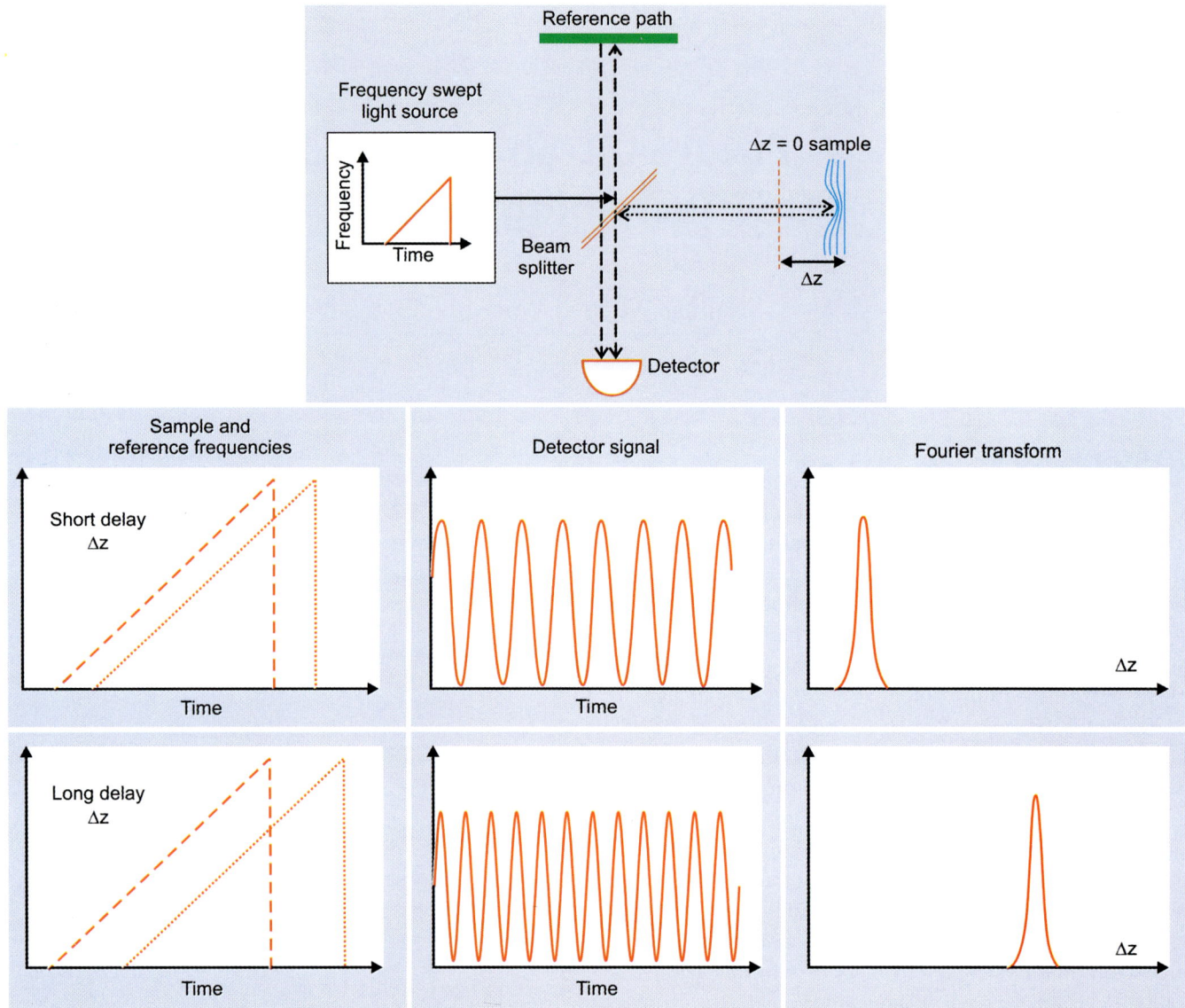

Fig. 11.1 How swept-source optical coherence tomography (SS-OCT) works

by Chinn et al.[2] and Golubovic et al.,[1] but performance was limited by available laser technology. Increases in SS-OCT imaging speed have been closely linked to advances in lasers, because axial scan rates are determined by the laser sweep repetition rate. In 2003, Yun et al. demonstrated OCT imaging with 19,000 axial scans per second and 13–14 um axial resolution (in air).[5] Imaging speeds of 115,000 axial scans per second were achieved using swept laser technology with a diffraction grating and rotating polygon mirror tuner in 2005.[6] The development of a new swept laser technology known as Fourier domain mode locking (FDML) (not related to FD-OCT) by Huber et al. in 2006 overcame fundamental limitations to laser sweep speed and enabled dramatic increases in imaging speed.[7] In 2008, Srinivasan et al. used FDML lasers to demonstrate retinal imaging at 249,000 axial scans per second and 1050 nm wavelength with an 8 um axial resolution.[8] Ultrahigh speed retinal imaging at 1,370,000 axial scans per second was demonstrated by Klein et al. in 2011 using FDML lasers, but the extremely high speeds required performance trade-offs and image resolution was 19 um and sensitivity was 92 dB.[9]

The availability and cost of swept laser technology is the major limiting factor for commercializing next generation SS-OCT. Commercially available swept lasers at 1,050 nm using polygon mirror tunable filter technology were employed for SS-OCT retinal imaging at 28,000 axial scans per second and 10 um axial resolution as early as 2007.[10] Commercial swept lasers using short cavities and microelectromechanical systems (MEMS) tunable filter technology could achieve higher imaging speeds of 100,000–200,000 axial scans per second with 7 um axial resolutions.[11] The most recent advances have been achieved by vertical-cavity surface-emitting swept laser (VCSEL) technology. Imaging speeds of 580,000 axial scans per second were recently demonstrated at 1,050 nm with 9 um axial resolution.[12] This chapter describes results using SS-OCT at 400,000 axial scans per second, factors of 4–5 times faster than current commercial OCT technology.

OPTICAL COHERENCE TOMOGRAPHY-ANGIOGRAPHY

As described in previous chapters, OCTA generates three-dimensional (3D) images of vascular structure without requiring exogenous dyes, such as fluorescein or indocyanine green. OCTA imaging protocols acquire multiple cross-sectional images (B-scans) from the same retinal position in order to detect changes which are produced by moving blood cells or flow. OCTA generates motion contrast. If the tissue is stationary, then all of the pixels in repeated B-scans will be the same. However, if there is motion from blood flow, then there will be fluctuations in the OCT intensity or phase at pixels where there are blood vessels. These fluctuations can be characterized by a decorrelation signal which is calculated at each pixel. Many techniques for OCTA using intensity and/or phase information have been described by several research groups.[13-21]

Figures 11.2A1 to C3 show an example of structural OCT and OCTA of the normal retina illustrating the role of ultrahigh speed in achieving wide field imaging. The images were obtained using prototype SS-OCT technology with a VCSEL light source at 400,000 axial scans per second with a 9.6 um axial resolution at 1,060 nm. OCTA was performed by repeatedly acquiring five B-scans from the same retinal position, each consisting of 500 A-scans with 500 different B-scan positions in the volumetric data set. Each OCTA data set required 500 × 5 × 500 A-scans and was acquired in 3.9 seconds. The figure shows en-face OCTA of over different fields ranging from 12 mm × 12 mm to 6 mm × 6 mm and 3 mm × 3 mm. Since OCTA generates 3D images of vasculature, both retinal and choroid vasculatures can be obtained from the same data set. The examples show *en-face* retinal vasculature summing the angiogram above the retinal pigment epithelium (RPE) and choriocapillaris/choroidal vasculature at a level approximately 20 um below the RPE. Note that each field of view differs by a factor of two in linear dimension or four in area. If future OCT instruments operate at 200,000–400,000 axial scans per second, approximately 2–4 times faster than current instruments, they will enable an approximately 1.4–2 times larger linear field to be imaged while maintaining the same resolution.

IMAGING THE CHORIOCAPILLARIS

Imaging the choroid and choriocapillaris is important for assessing early markers of retinal disease. Choroidal circulation supplies oxygen to the outer retinal layers and alterations are associated with posterior segment diseases, such as age-related macular degeneration (AMD) and diabetic retinopathy (DR).[22-24] The choriocapillaris is the inner layer of the choroid which supports RPE and photoreceptor metabolism. Imaging the choriocapillaris is especially challenging because it is located behind the pigmented RPE; is quite thin, even in healthy eyes; and has dense lobular microvasculature which requires extremely high pixel densities to resolve. Previous studies using *en-face* OCT structural imaging have investigated the choriocapillaris. Motaghiannezam et al. demonstrated visualization of the choriocapillaris and larger choroidal vessels in the Sattler's and Haller's layers in normal subjects using SS-OCT at the 57,000 axial scans per second.[25] Sohrab et al. also studied the choriocapillaris and larger choroidal vessel patterns with SD-OCT in patients with early AMD or reticular pseudodrusen.[26] While structural OCT imaging methods for the choriocapillaris can be clinically useful, OCTA can be more sensitive for visualizing alterations in vascular structure.

Optical coherence tomography-angiography of the choriocapillaris in normal subjects was demonstrated using adaptive optics and Doppler phase techniques with long wavelength SD-OCT at 91,000 axial scans per second by Kurokawa et al. in 2012.[27] OCTA of the choriocapillaris in a normal subject was also demonstrated using long wavelength SS-OCT at 100,000 axial scans per second combined with scanning laser ophthalmoscope (SLO) eye tracking by Braaf et al. in 2013.[28] Studies in an AMD patient and a normal subject were reported by Kim et al. in 2013 using phase variance OCTA of the choriocapillaris with SD-OCT.[29]

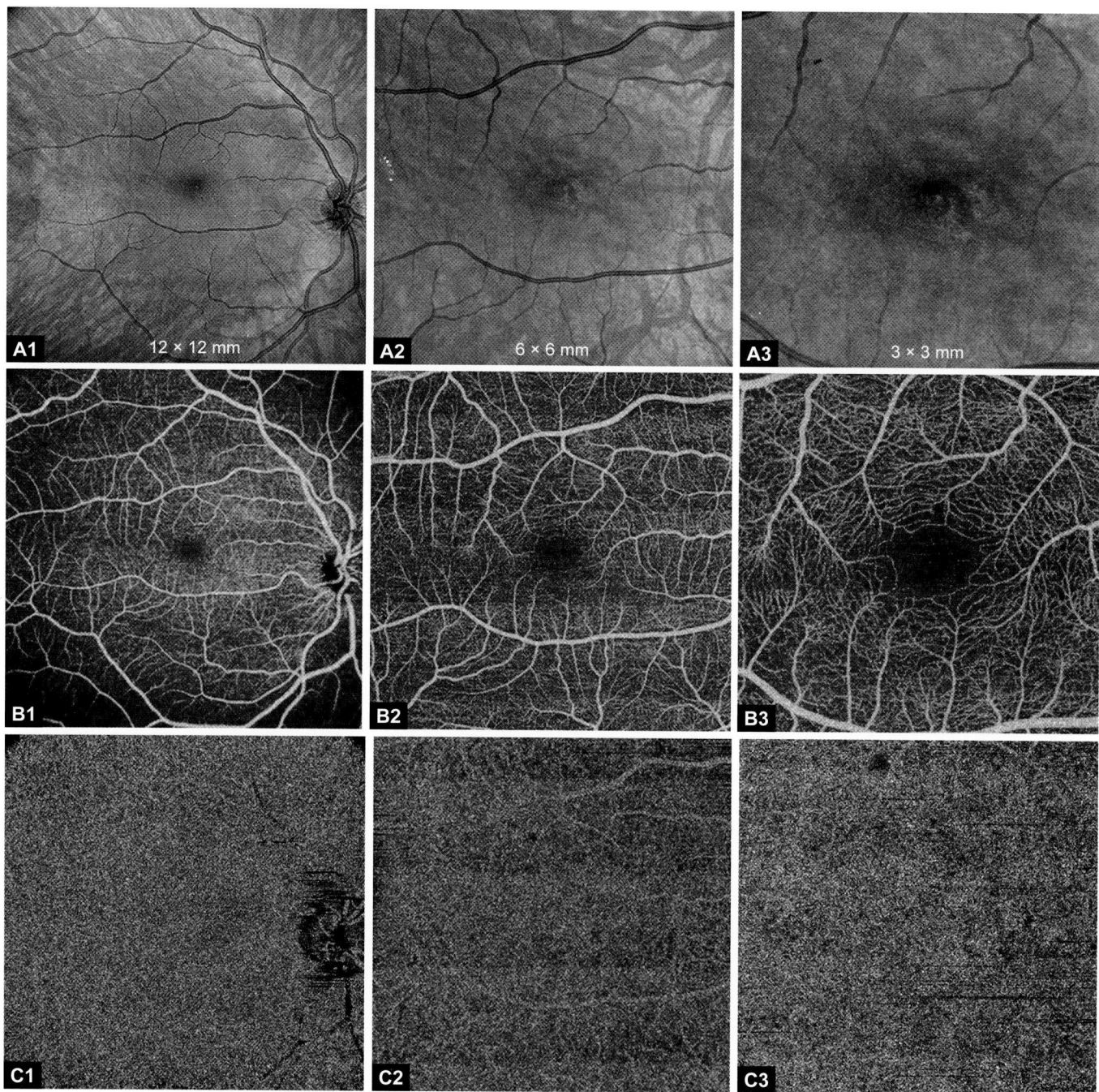

Figs 11.2A1 to C3 Ultrahigh speed enables wide field optical coherence tomography (OCT) angiography (OCTA): OCT structural (top row; A1 to A3) and angiography images of the retina (middle row; B1 to B3) and choriocapillaris/choroid (bottom row; C1 to C3). As the scanning area is reduced from 12 mm × 12 mm to 6 mm × 6 mm and 3 mm × 3 mm, there is an increase in the resolution and finer vascular detail can be visualized

These studies were all performed with acquisition speeds of approximately 100,000 axial scans per second and used relatively long imaging times or small imaging areas.

Figure 11.3 shows *en-face* OCT intensity and OCTA images of the choriocapillaris of a normal subject generated by mosaicking smaller 3 mm × 3 mm fields over a 32 mm field

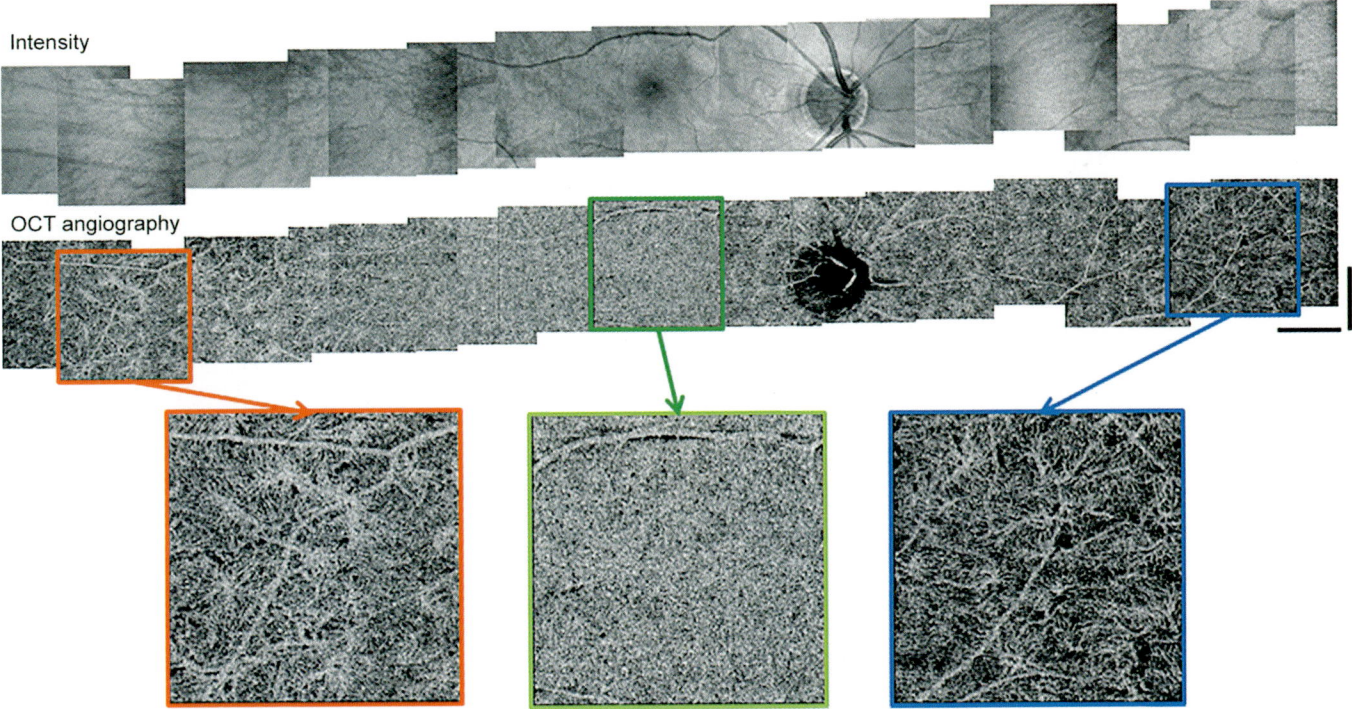

Fig. 11.3 Ultrahigh speed and long wavelength enable choriocapillaris imaging. The choriocapillaris is challenging to image because of its fine vascular structure. This example shows mosaic imaging with small regions at various fundus locations from the macula to the periphery. Note the lobular organization of the choriocapillaris vasculature in the periphery which becomes finer in the macular region
Source: Adapted and reprinted with permission from reference 30

spanning the macula and peripheral retina from Choi et al.[30] Imaging was performed using a prototype SS-OCT instrument using a VCSEL light source operating at 400,000 axial scans per second with a 9.6 um axial resolution at 1,060 nm. OCTA was performed by repeatedly acquiring four B-scans from the same retinal position, each consisting of 800 A-scans with 400 different B-scan positions in the volumetric data set. Each OCTA data set required 800 × 4 × 400 A-scans and was acquired in 3.8 seconds.

The *en-face* OCT angiograms shown were at a depth immediately below and adjacent to the RPE. The OCTA findings agree with known histological and electron micrograph corrosion casting studies which showed that the capillary density and pattern in the choriocapillaris changes depending on the fundus location, with a densely packed honeycomb structure at the central fovea, and a more lobular and less dense structure towards the equator and periphery.[31-34] Because of the extremely small vascular structure in the choriocapillaris, ultrahigh speed imaging is required to achieve the high pixel densities needed to resolve these features in *en-face* OCTA.

OPTICAL COHERENCE TOMOGRAPHY-ANGIOGRAPHY IN DIABETES

Optical coherence tomography-angiography is especially useful in retinal vascular diseases, where it enables visualization of the retinal microvasculature. Figures 11.4A1 to A3 compares a fundus photograph, fluorescein angiography (FA) and ultrahigh speed SS-OCTA of a 45-year-old male patient with DR. The fluorescein angiogram exhibits multiple small microaneurysms (Figs 11.4B1 and B2), vascular remodeling and enlargement of the foveal avascular zone. All features are visualized at least as well and possibly better in the OCT angiogram of the retinal vasculature, which shows vascular pruning and remodeling, enlargement and irregularity of the foveal avascular zone, and capillary dropout, especially temporal to the foveal avascular zone. The

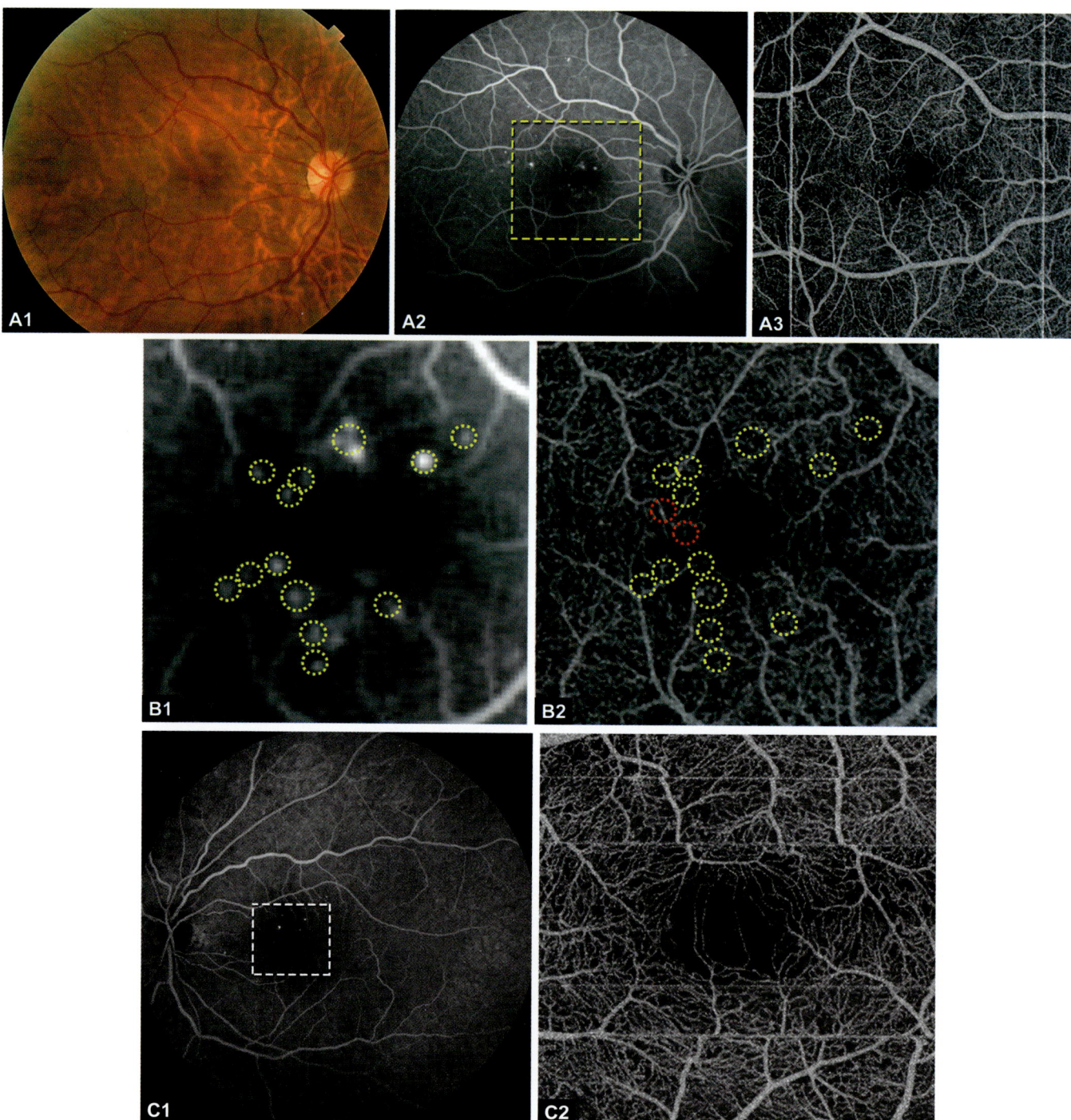

Figs 11.4A1 to C2 (A1 to A3) Moderate nonproliferative diabetic retinopathy (NPDR) with diabetic macular edema (DME). Optical coherence tomography (OCT) angiography (OCTA) of a 45-year-old man with moderate NPDR and DME. Note the exquisite retinal vascular detail visible on the OCT angiogram on the right; (B1 and B2) Microaneurysms on fluorescein angiography (FA) and OCTA. FA can identify leaking microaneurysms in the macula (yellow circles). OCTA shows that most of the leaking microaneurysms are located adjacent to areas of capillary nonperfusion, a detail that cannot be appreciated in the fluorescein angiogram. Moreover, OCTA also identifies non-leaking aneurysms (examples outlined with red circles); (C1 and C2) FA and OCTA of the contralateral eye. OCTA of the contralateral eye of the same patient shows the same level of detail as visualized in the right eye. Since OCTA does not require dye, there is no "acute" and "late" eye on the OCT angiogram. Vasculature in both eyes can be visualized in equal detail

additional advantage of OCTA is that because fluorescein is not required, the imaging results from the "acute" eye are the same as "late" or "intermediate" eye; OCTA of both eyes can be performed with equal detail. Figures 11.4C1 and C2 shows an OCTA of the contralateral eye.

Optical coherence tomography-angiography has the limitation that it cannot visualize alterations in vascular permeability, whereas FA shows dye leakage from abnormal retinal blood vessels. However, hyperfluorescence from dye leakage in FA can also be a disadvantage because it can obscure fine vascular structure. In addition, since structural OCT images are acquired simultaneously with OCTA, a retinal thickness map that is precisely registered to the angiogram can be generated, identifying areas of retinal thickening that can be used as a surrogate for leakage. When coupled with the ability to visualize microaneurysms on OCTA, structural OCT thickness maps can be a useful guide for application of focal laser, one of the most important clinical indications for mapping fluorescein dye leakage.

In addition, simultaneous choriocapillaris angiograms can also be obtained on these patients. Figures 11.5A and B show an OCT angiogram from a 68-year old male patient with a history of noninsulin-dependent diabetes mellitus (DM) without clinically evident DR, exhibiting patchy loss or reduction in flow in the choriocapillaris. Our studies indicate that patchy or diffuse loss or flow reduction in the choriocapillaris is present in many patients with diabetes, even in the absence of DR. This correlates with histopathological observations made in the eyes of diabetics with no clinical evidence of DR.[24]

OPTICAL COHERENCE TOMOGRAPHY-ANGIOGRAPHY IN DRY AGE-RELATED MACULAR DEGENERATION

Optical coherence tomography-angiography promises to be especially useful in AMD since this disease involves the choriocapillaris/choroidal vasculature, and the choriocapillaris is especially difficult to image in vivo using any other imaging modality. Current generation OCTA using commercial instruments can image both classic and occult choroidal neovascularization. However, OCTA using long wavelength, ultrahigh speed SS-OCT enables imaging fine detail in the choriocapillaris. This may enable researchers to identify structural or flow markers that predict progression of macular degeneration, as well as to better understand the pathogenesis of disease. Figures 11.6A to C show images from a 75-year-old patient with a history of geographic atrophy (GA), secondary to AMD. SS-OCTA studies on AMD patients with GA show complete choriocapillaris degeneration in the area of the GA, as shown in Figure 6. However, alterations in the choriocapillaris often extend beyond the margin of the GA and may be related either to loss of vasculature or reduced flow in the choriocapillaris.

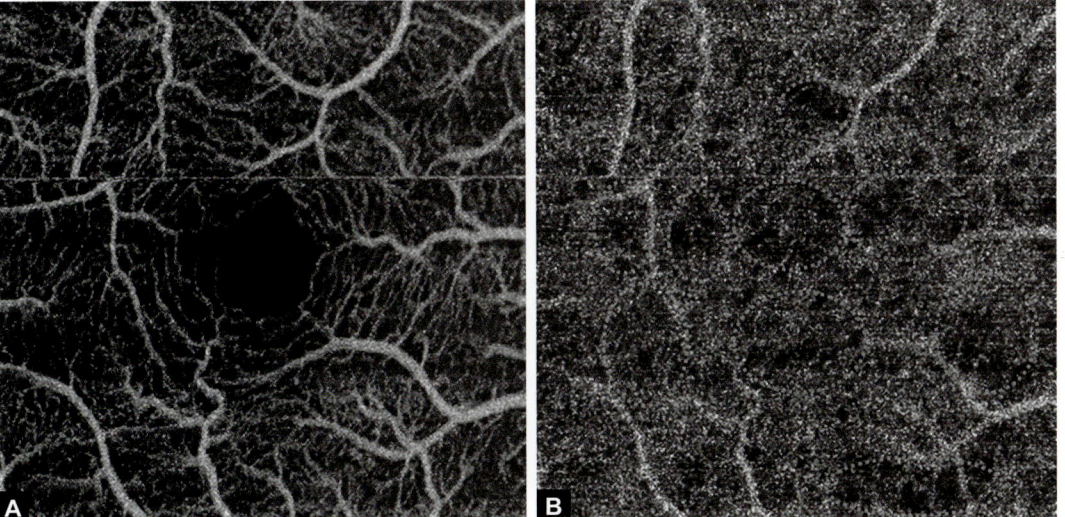

Figs 11.5A and B Diabetes mellitus (DM) without diabetic retinopathy (DR): Optical coherence tomography-angiography (OCTA) of a 68-year-old man with a history of DM, but with no clinically evident DR. OCTA shows normal retinal vasculature, but patchy loss or flow reduction in the choriocapillaris

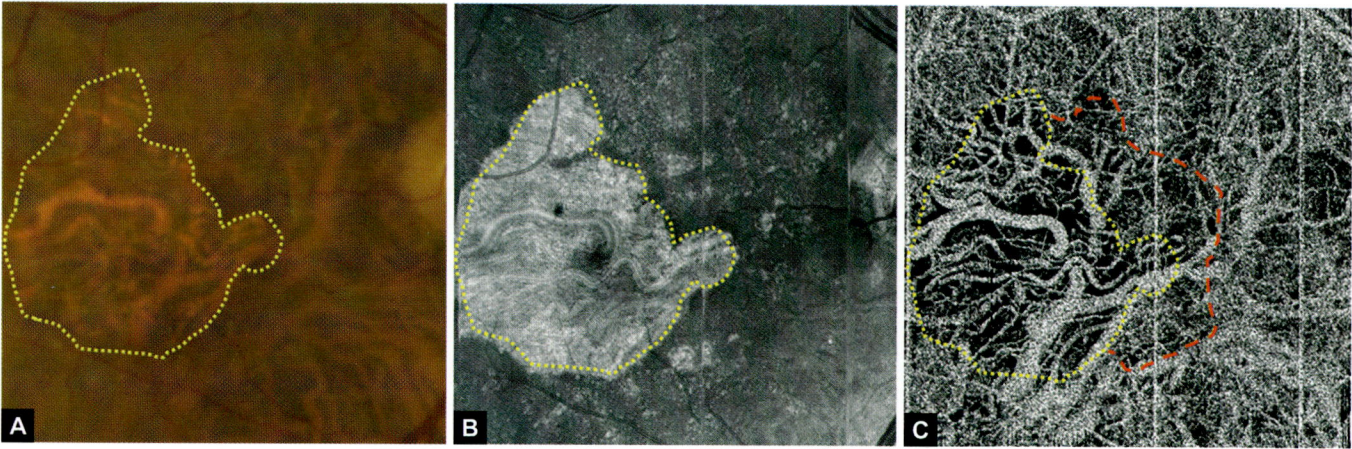

Figs 11.6A to C Geographic atrophy (GA) in dry age-related macular degeneration (AMD): A 75-year-old patient with AMD and GA. The *en-face* optical coherence tomography (OCT) intensity projection from a sub-retinal pigment epithelium (RPE) slab (left) clearly delineates the region of GA. OCT-angiography (OCTA) of the choriocapillaris (right) exhibits complete atrophy in the region of GA (yellow outline). However, OCTA also shows flow reduction or atrophy in a region extending outside the GA (red outline). These features may be early markers for progression or response to therapy

SUMMARY

Optical coherence tomography-angiography is a powerful method for visualizing retinal and choroidal vasculature. Next generation SS-OCT technology promises to enable wide field OCTA that simultaneously provides exquisite detail of retinal as well as choriocapillaris microvasculature. Since 3D structural OCT images are acquired simultaneously with, and are precisely registered to OCTA, this provides comprehensive information about pathology. OCTA is noninvasive, without the injection of dye with its concomitant risks of extravasation and anaphylaxis, and repeated imaging can be performed on successive patient visits to assess progression or response. As OCTA becomes more widely available and comprehensive clinical information emerges, it promises to have a significant impact on pharmaceutical development, understanding disease pathogenesis as well as clinical practice.

At the same time, OCTA is an extremely demanding imaging modality because protocols require repeated scanning of the same retinal position and each *en-face* OCTA image pixel requires multiple axial scans. Since area increases as the square of the linear dimension, there are severe demands on imaging speeds to achieve wider retinal coverage. The development of new ultrahigh speed SS-OCT promises to increase imaging speeds by approximately 4–5 times compared with current clinical technology. The transition of new technology from research prototype to widespread clinical practice can be time consuming because of complex engineering and regulatory requirements. However, the next generation of ultrahigh speed technology promises to achieve even further improvements in OCT functionality, retinal coverage and image quality, enabling more sensitive diagnosis, assessment of disease progression and response to therapy.

ACKNOWLEDGMENTS

The authors gratefully acknowledge Mehreen Adhi and Eric Moult for assistance with imaging and image processing, Jonathan Liu, Chen Lu for developing the swept source instrument, Alex Cable from Thorlabs and Vijaysekhar Jayaraman from Praevium Research for developing VCSEL laser technology. We also gratefully acknowledge support from the National Institutes of Health (NIH R01-EY011289-27, R44 EY022864-01), Air Force Office of Scientific Research (AFOSR FA9550-10-1-0063 and AFOSR FA9550-10-1-0063), the Champalimaud Vision Award Fund, and an unrestricted grant from Research to Prevent Blindness and the Massachusetts Lions Clubs.

REFERENCES

1. Golubovic B, Bouma BE, Tearney GJ, Fujimoto JG. Optical frequency-domain reflectometry using rapid wavelength tuning of a Cr4+:forsterite laser. Opt Lett. 1997;22(22):1704-6.
2. Chinn SR, Swanson EA, Fujimoto JG. Optical coherence tomography using a frequency-tunable optical source. Opt Lett. 1997;22(5):340-2.
3. Unterhuber A, Povazay B, Hermann B, Sattmann H, Chavez-Pirson A, Drexler W. In vivo retinal optical coherence tomography at 1040 nm - enhanced penetration into the choroid. Opt Express. 2005;13(9):3252-8.
4. Povazay B, Hermann B, Unterhuber A, Hofer B, Sattmann H, Zeiler F, et al. Three-dimensional optical coherence tomography at 1050 nm versus 800 nm in retinal pathologies: enhanced performance and choroidal penetration in cataract patients. J Biomed Opt. 2007;12(4):041211.
5. Yun SH, Tearney GJ, Bouma BE, Park BH, de Boer JF. High-speed spectral-domain optical coherence tomography at 1.3 μm wavelength. Opt Express. 2003;11(26):3598-604.
6. Oh WY, Yun SH, Tearney GJ, Bouma BE. 115 kHz tuning repetition rate ultrahigh-speed wavelength-swept semiconductor laser. Opt Lett. 2005;30(23):3159-61.
7. Huber R, Wojtkowski M, Fujimoto JG. Fourier Domain Mode Locking (FDML): A new laser operating regime and applications for optical coherence tomography. Opt Express. 2006;14(8):3225-37.
8. Srinivasan VJ, Monson BK, Wojtkowski M, Bilonick RA, Gorczynska I, Chen R, et al. Characterization of outer retinal morphology with high-speed, ultrahigh-resolution optical coherence tomography. Invest Ophthalmol Vis Sci. 2008;49(4):1571-9.
9. Klein T, Wieser W, Eigenwillig CM, Biedermann BR, Huber R. Megahertz OCT for ultrawide-field retinal imaging with a 1050 nm Fourier domain mode-locked laser. Opt Express. 2011;19(4):3044-62.
10. Yasuno Y, Hong YJ, Makita S, Yamanari M, Akiba M, Miura M, et al. In vivo high-contrast imaging of deep posterior eye by 1-μm swept source optical coherence tomography and scattering optical coherence angiography. Opt Express. 2007;15(10):6121-39.
11. Potsaid B, Baumann B, Huang D, Barry S, Cable AE, Schuman JS, et al. Ultrahigh speed 1050 nm swept source/Fourier domain OCT retinal and anterior segment imaging at 100,000 to 400,000 axial scans per second. Opt Express. 2010;18(19):20029-48.
12. Grulkowski I, Liu JJ, Potsaid B, Jayaraman V, Lu CD, Jiang J, et al. Retinal, anterior segment and full eye imaging using ultrahigh speed swept source OCT with vertical-cavity surface emitting lasers. Biomed Opt Express. 2012;3(11):2733-51.
13. Makita S, Hong Y, Yamanari M, Yatagai T, Yasuno Y. Optical coherence angiography. Opt Express. 2006;14(17):7821-40.
14. Fingler J, Schwartz D, Yang CH, Fraser SE. Mobility and transverse flow visualization using phase variance contrast with spectral domain optical coherence tomography. Opt Express. 2007;15(20):12636-53.
15. Tao YK, Kennedy KM, Izatt JA. Velocity-resolved 3D retinal microvessel imaging using single-pass flow imaging spectral domain optical coherence tomography. Opt Express. 2009;17(5):4177-88.
16. An L, Wang RK. In vivo volumetric imaging of vascular perfusion within human retina and choroids with optical micro-angiography. Opt Express. 2008;16(15):11438-52.
17. Mariampillai A, Standish BA, Moriyama EH, Khurana M, Munce NR, Leung MK, et al. Speckle variance detection of microvasculature using swept-source optical coherence tomography. Opt Lett. 2008;33(13):1530-2.
18. Vakoc BJ, Lanning RM, Tyrrell JA, Padera TP, Bartlett LA, Stylianopoulos T, et al. Three-dimensional microscopy of the tumor microenvironment in vivo using optical frequency domain imaging. Nat Med. 2009;15(10):1219-23.
19. Yu LF, Chen ZP. Doppler variance imaging for three-dimensional retina and choroid angiography. J Biomed Opt. 2010;15(1):016029.
20. Enfield J, Jonathan E, Leahy M. In vivo imaging of the microcirculation of the volar forearm using correlation mapping optical coherence tomography (cmOCT). Biomed Opt Express. 2011;2(5):1184-93.
21. Blatter C, Klein T, Grajciar B, Schmoll T, Wieser W, Andre R, et al. Ultrahigh-speed non-invasive widefield angiography. J Biomed Opt. 2012;17(7):070505.
22. Schmetterer L, Kiel JW (Eds). Ocular Blood Flow. Berlin, Heidelberg: Springer-Verlag; 2012.
23. Lutty GA, Cao JT, McLeod DS. Relationship of polymorphonuclear leukocytes to capillary dropout in the human diabetic choroid. Am J Pathol. 1997;151(3):707-14.
24. Cao JT, McLeod DS, Merges CA, Lutty GA. Choriocapillaris degeneration and related pathologic changes in human diabetic eyes. Arch Ophthalmol. 1998;116(5):589-97.
25. Motaghiannezam R, Schwartz DM, Fraser SE. In vivo human choroidal vascular pattern visualization using high-speed

swept-source optical coherence tomography at 1060 nm. Invest Ophthalmol Vis Sci. 2012;53(4):2337-48.
26. Sohrab M, Wu K, Fawzi AA. A pilot study of morphometric analysis of choroidal vasculature in vivo, using en face optical coherence tomography. PloS One. 2012;7(11): e48631.
27. Kurokawa K, Sasaki K, Makita S, Hong YJ, Yasuno Y. Three-dimensional retinal and choroidal capillary imaging by power Doppler optical coherence angiography with adaptive optics. Opt Express. 2012;20(20):22796-812.
28. Braaf B, Vienola KV, Sheehy CK, Yang Q, Vermeer KA, Tiruveedhula P, et al. Real-time eye motion correction in phase-resolved OCT angiography with tracking SLO. Biomed Opt Express. 2013;4(1):51-65.
29. Kim DY, Fingler J, Zawadzki RJ, Park SS, Morse LS, Schwartz DM, et al. Optical imaging of the chorioretinal vasculature in the living human eye. Proc Natl Acad Sci U S A. 2013;110(35): 14354-9.
30. Choi W, Mohler KJ, Potsaid B, Lu CD, Liu JJ, Jayaraman V, et al. Choriocapillaris and choroidal microvasculature imaging with ultrahigh speed OCT angiography. PloS One. 2013;8(12):e81499.
31. Yoneya S, Tso MO. Angioarchitecture of the human choroid. Arch Ophthalmol. 1987;105(5):681-7.
32. Olver JM. Functional anatomy of the choroidal circulation: methyl methacrylate casting of human choroid. Eye (Lond). 1990;4(Pt 2):262-72.
33. McLeod DS, Lutty GA. High-resolution histological analysis of the human choroidal vasculature. Invest Ophthalmol Vis Sci. 1994;35(11):3799-811.
34. Zhang HR. Scanning electron-microscopic study of corrosion casts on retinal and choroidal angioarchitecture in man and animals. Prog Retinal Eye Res. 1994;13(1):243-70.

Index

Page numbers followed by f refer to figure and t refer to table

A
Age-related
  macular degeneration 21, 49f-55f, 59f, 60, 73f, 77, 82f
  retinal anomalies 23
Angiomatoses 3, 27f
Angio-OCT 1, 2t, 3t, 23, 32f, 38f, 40f, 41f, 54f, 55f, 68, 71
  advantages 69
  clinical terminology 5
  disadvantages 69
  signal 17
Anti-vascular endothelial growth factor 22, 52f, 73f
Arteries 12
  occlusions 19
Atrophic macular degeneration 59
Atrophy, geographic 59f, 82f

B
Blood disorders 19
Branch retinal vein occlusion 30, 32f
Branch vein occlusion 33f-36f
  of superotemporal
    retina of fovea 35f
    vein of fovea 36f
Break scar, choroidal 54f
Bruch's membrane 5, 47, 48f, 49f, 58f, 71-73

C
Capillaries
  density of 20t
  lesions 20
  microocclusions of 21
  network, meshes of 20t
Choroid 71
Choroiditis 3
Classical membrane, neovascular 48f
Coat's disease 23, 24f-27f
Cystoid edema cells, small 24f, 42f, 43f

D
Decorrelation signal
  intensity of 18
  vascular 6, 17

Deep vascular plexus 13, 31f, 39f, 42f, 44f, 71, 28f, 72
Diabetes mellitus 81f
Diabetic retinopathy
  ophthalmoscopic 72f
  proliferative 45, 46f, 74f

E
Eales' disorder 19
Edema 2, 19, 21, 26f
  corneal 2
  cystoid 27f, 30f
Endothelial growth factor, vascular 7
Epithelial disorders
  acute 1
  chronic 1
Eye disorders 23

F
Fibrosis 48
Fluorangiography 3f, 3t, 32f, 35f, 39f, 68, 68f
  retinal 3
Fluorescein angiography 10, 55f, 68, 69

G
General fluorangiography
  advantages 68
  disadvantages 68
Glaucoma 2, 3

H
Hemorrhage
  retinal 17
  vitreal 2

I
Idiopathic polypoidal choroidal vasculopathy 49, 58f
Inflammations
  choroidal 3
  retinal 3
Ischemia 21
  retinal 32, 35
Ischemic diabetic retinopathy, intraretinal layers of 43f

J
Juvenile macular degenerations 2

L
Lens opacities 2
Limiting membrane, internal 11, 13, 13f, 14f, 16

M
Macroaneurysm 25, 29f, 30f
Macular
  edema, diabetic 80f
  new vessels, subretinal 3
Maculopathy, ischemic 39f, 40f
Malformations, vascular 3
Membrane, neovascular 47, 49f, 73f
Microelectromechanical systems 77
Microvascularization, intraretinal 31f

N
Neovascular membrane
  analysis of 21
  classical 48f
  in myopic eyes 48
  myopic 56f, 57f, 58f
  occult 50f
  subretinal 51f
Neovascular proliferation, preretinal 3
Neovascularization
  choroidal 7, 48f, 49f, 60, 62f, 69, 71, 72
  intraretinal 2
  preretinal 2
  subretinal 1
Nerve fiber layer 64
Nodular subretinal membranes, neovascular 55f
Nonproliferative diabetic retinopathy 31f, 37f-40f, 42f, 43f, 80f
Normal retina, anatomy of 15f

O
Occlusions, vascular 3
Occult membrane, neovascular 48f, 49f
Opacities, corneal 2

Optic disk
- disorders 2, 3, 59
- normal 59f

Optical coherence tomography 8, 9f, 13f, 60, 64, 65f, 75, 82f
- angiography 1, 2, 2t, 3, 3t, 5, 10, 11, 14, 15f, 16, 23, 24f, 30f-34f, 37f, 38f, 40f, 41f, 49, 54f, 55f, 58f, 60, 61f, 66f, 68, 69, 71, 71f, 75, 77, 78f, 80f, 81f
  - advantages 68f
  - disadvantages 68
  - in diabetes 79
  - in dry age-related macular degeneration 81
  - of normal retina 12
  - of optic disk and peripapillary retinal perfusion in glaucoma 64
  - principles of 1
- B-scan 13f
- limitations of 2

P

Perifoveal capillary network in angio-OCT, anomalies of 19

Pigment epithelium
- retinal 5, 11, 16, 71-73
- subretinal 82f

Pigment
- accumulation of 17
- epithelium vascularized detachment 48f, 49f

Plexiform layer
- inner 11, 13, 14f
- outer 11, 14f, 60

Plexus
- superficial 13, 13f, 15, 15f, 25f, 24f
- vascular 13f

Prepapillary membrane, neovascular 46f

Pressure, intraocular 64

Pucker, macular 29f

Q

Quantification of
- optic disk perfusion 64
- peripapillary retinal perfusion 64

R

Retina
- inner 23
- normal 14f
- outer 47
- pigment epithelium 47

Retinal vascular
- networks, retinal 12
- plexuses with angio-OCT, analysis of 12

Retinal vessels, superficial 14f

Retinopathies
- acquired 1, 3
- background 35
- congenital 1, 3
- diabetic 1, 3, 19, 35, 38f, 41f, 42f, 43f, 71f, 73, 81f
- general 3
- ischemic 19

S

Segmentation, thickness of 7

Sickle cell anemia 19

Spectral optical coherence tomography 64

Split-spectrum amplitude-decorrelation angiography 1, 8, 11, 17, 35, 60, 62f, 68, 69, 71

Subretinal membranes, neovascular 55f

Superficial plexus, evaluation of 13f

T

Telangiectasias, macular 28f

Thalassemia 19

U

Uveitis 3

V

Vascular
- network of retina, superficial 29f
- plexus, superficial 13, 14f, 28f, 31f, 36f, 37f, 38f, 42f, 71, 72

Vasculitis 19

Vein
- occlusion 19
- retinal 12

Vessels
- aspects of 18t
- shape of 20t
- size of 19t

Vitreous
- hazy 2
- opaque 2